D1079996

MODULAR MATHEMATICS

Module A: Pure Maths 1

Second Edition

By the same authors

MODULE B: PURE MATHS 2
MODULE C: STATISTICS 1
MODULE D: STATISTICS 2 (due 1996)
MODULE E: MECHANICS 1
MODULE F: MECHANICS 2
CORE MATHS FOR A-LEVEL
MATHEMATICS — THE CORE COURSE FOR A-LEVEL
FURTHER PURE MATHEMATICS — with C. Rourke
MATHEMATICS — MECHANICS AND PROBABILITY
FURTHER MECHANICS AND PROBABILITY
APPLIED MATHEMATICS I
APPLIED MATHEMATICS II
PURE MATHEMATICS I
PURE MATHEMATICS II

MODULAR MATHEMATICS

Module A: Pure Maths 1

Second Edition

L. Bostock, B.Sc.

S. Chandler, B.Sc.

Stanley Thornes (Publishers) Ltd

90545

First published in 1990
Second edition published in 1995 by:
Stanley Thornes (Publishers) Ltd
Ellenborough House
Wellington Street
CHELTENHAM GL50 1YW

97 98 99 00 01 / 10 9 8 7 6 5 4 3

A catalogue record of this book is available from the British Library.

ISBN 0-7487-1777-3

Cover photograph: the computer program *Mathematica* featured courtesy of Wolfram Research

Typeset by Tech-Set, Gateshead, Tyne & Wear.
Printed and bound in Great Britain at TJ Press, Padstow, Cornwall

CONTENTS

PREFACE TO THE SECOND EDITION

This second edition of Module A has been extensively revised to cover the work required for the P1 syllabus of the London Modular Mathematics which has itself been changed to comply with the new subject core for A- and AS-level.

This book assumes, as a starting point, the minimum level of achievement on the national curriculum for access to A-level. This currently stands at level 7/8. Many of you will have reached a higher level and so will find an overlap between some of the work in this book and what you already know. Chapters containing such topics do, however, cover some questions that are more sophisticated than those likely to be met at Key Stage 4. Before assuming that you can tackle these more demanding problems, we suggest that you use the mixed exercises at the ends of these chapters, except for the starred questions, to identify any weaknesses. Further work to deal with any troublespots can then be carried out using the relevant sections of the chapter.

The starred questions at the ends of some chapters are not part of the minimum requirement for this syllabus. Ideas needed in some of these questions extend beyond the scope of the syllabus. Others illustrate some interesting, but more difficult, applications of the topic while others are simply harder problems. They are there to stretch those of you who like being challenged.

The exercises in the chapters all start with straightforward questions designed to build up confidence. Some of them are fairly long and appear repetitive; this is because much practice is needed for algebraic processes to become instinctive in much the same way as working with numbers should be (if you even think about getting your calculator out to evaluate 200×0.7, say, you should do some extensive revision of number work). More sophisticated questions of the examination type are given in consolidation sections that appear at regular intervals throughout the book. These are intended for use later, to give practice when confidence has been built up. The consolidation sections also include a summary of the work in preceding chapters and a set of multiple choice questions which probe understanding in a way that other questions cannot do.

There are many computer programs that aid the understanding of mathematics. In particular, a good graph-drawing package is invaluable for investigating graphical aspects of functions. Graphics calculators are also invaluable and some of them now available are capable of numerical solution of equations and numerical differentiation and integration. In a few places we have indicated where such aids can be used effectively, but this should be regarded as a minimum indication of the use of technology. Some computer programs, such as *Mathematica*, are capable of handling a wide variety of symbolic algebraic processes. Two publications that we recommend in this connection are: *Graphics Calculators in the Mathematics Classroom* and *Spreadsheets: Exploring their Potential in Secondary Mathematics*. These are published by The Mathematical Association and are available from the Association at 259 London Road, Leicester, LE2 3BE, UK.

We are grateful to the following examination boards for permission to reproduce questions from their past examination papers. Part questions are indicated by the suffix p. Questions from specimen papers are indicated by the suffix s, and it should be noted that these questions have not been subjected to the rigorous checking and moderation procedure by the Boards that their examination questions undergo.

University of London Examinations and Assessment Council (ULEAC)
The Associated Examining Board (AEB)
University of Cambridge Local Examinations Syndicate (UCLES)
Northern Examinations and Assessment Council (NEAB)
Welsh Joint Education Committee (WJEC)
Oxford and Cambridge Schools Examination Board (OCSEB, MEI)

L. Bostock
S. Chandler
1995

USEFUL INFORMATION

NOTATION

$=$	is equal to	$\in$	is a member of
$\equiv$	is identical to	$:$	is such that
$\approx$	is approximately equal to*	$\mathbb{N}$	the natural numbers
$>$	is greater than	$\mathbb{Z}$	the integers
$\geqslant$	is greater than or equal to	$\mathbb{Q}$	the rational numbers
$<$	is less than	$\mathbb{R}$	the real numbers
$\leqslant$	is less than or equal to	$\mathbb{R}^+$	the positive real numbers
∞	infinity; infinitely large		excluding zero
$\Rightarrow$	implies	$[a, b]$	the interval $\{x : a \leqslant x \leqslant b\}$
		$|x|$	the modulus of x

A stroke through a symbol negates it, e.g. $\neq$ means 'is not equal to'

ABBREVIATIONS

$\|$	is parallel to	$-$ve	negative
$+$ve	positive	w.r.t.	with respect to
		abs	absolute value of

USEFUL FORMULAE

For a cone with base radius r, height h and slant height l

$$\text{volume} = \tfrac{1}{3}\pi r^2 h \qquad \text{curved surface area} = \pi r l$$

For a sphere of radius r

$$\text{volume} = \tfrac{4}{3}\pi r^3 h \qquad \text{surface area} = 4\pi r^2$$

For any pyramid with height h and base area a

$$\text{volume} = \tfrac{1}{3}ah$$

COMPUTER AND CALCULATOR REFERENCES

Where computer programs or calculators can be helpful they are identified by marginal symbols in the following way

*Practical problems rarely have exact answers. Where numerical answers are given they are correct to two or three decimal places depending on their context, e.g. π is 3.142 correct to 3 d.p. and although we write $\pi = 3.142$ it is understood that this is not an exact value. We reserve the symbol $\approx$ for those cases where the approximation being made is part of the method used.

INSTRUCTIONS FOR ANSWERING MULTIPLE CHOICE EXERCISES

These exercises are included in each consolidation section. The questions are set in groups; the answering techniques are different for each group and are classified as follows:

TYPE I

These questions consist of a problem followed by several alternative answers, only *one* of which is correct.

Write down the letter corresponding to the correct answer.

TYPE II

A single statement is made. Write T if it is true and F if it is false.

CHAPTER 1

ALGEBRA

INTRODUCTION

Algebra is the use of letters and symbols to express generalisations. For example, we know that $3 + 3 = 2 \times 3$ and that $5 + 5 = 2 \times 5$; these are particular examples of the fact that any number added to itself is the same as twice that number. Using a letter to represent *any* number, this general fact can be written very concisely as $x + x = 2x$.

A general fact is much more powerful than particular facts; if we *only* knew that $3 + 3 + 3 + 3 = 4 \times 3$ we would not be able to say that $27 + 27 + 27 + 27$ is equal to 4×27; it is knowing that $x + x + x + x = 4x$ that enables us to write $27 + 27 + 27 + 27 = 4 \times 27$.

Mathematics beyond GCSE involves developing general methods that can then be used in a variety of ways; the ability to manipulate algebraic expressions is an essential base for any mathematics course beyond GCSE. Applying the processes involved needs to be almost as instinctive as the ability to manipulate simple numbers. This and the next two chapters present the facts and provide practice necessary for the development of these skills.

MULTIPLICATION OF ALGEBRAIC EXPRESSIONS

The multiplication sign is usually omitted, so that, for example,

$2q$ means $2 \times q$

and $x \times y$ can be simplified to xy

Remember also that if a string of numbers and letters are multiplied, the multiplication can be done in any order, for example

$$2p \times 3q = 2 \times p \times 3 \times q$$

$$= 6pq$$

Powers can be used to simplify expressions such as $x \times x$,

i.e. $x \times x = x^2$

and $x \times x^2 = x \times x \times x = x^3$

But remember that a power refers only to the number or letter it is written above, for example

$2x^2$ means that x is squared, but 2 is not

Example 1a

Simplify $(4pq)^2 \times 5$

$$(4pq)^2 \times 5 = 4pq \times 4pq \times 5$$
$$= 80p^2q^2$$

EXERCISE 1a

Simplify

1. $3 \times 5x$

2. $x \times 2x$

3. $(2x)^2$

4. $5p \times 2q$

5. $4x \times 2x$

6. $2pq \times 5pr$

7. $(3a)^2$

8. $7a \times 9b$

9. $8t \times 3st$

10. $2a^2 \times 4a$

11. $b^2 \times 4ab$

12. $(7pq)^2 \times (2p)^2$

13. $(2ab)^2 \times 3b$

14. $(5x)^2 \times (2y)^2$

15. $(2a)^2b \times 3ba^2$

ADDITION AND SUBTRACTION OF EXPRESSIONS

The *terms* in an algebraic expression are the parts separated by a plus or minus sign.

Like terms contain the same combination of letters; like terms can be added or subtracted.

For example, $2ab$ and $5ab$ are like terms and can be added,

i.e. $2ab + 5ab = 7ab$

Unlike terms contain different algebraic expressions; they cannot be added or subtracted. For example, ab and ac are unlike terms and $ab + ac$ cannot be collected together.

Example 1b

Simplify $5x - 3(4 - x)$

$$5x - 3(4 - x) = 5x - 12 + 3x$$
$$= 8x - 12$$

Note that $-3(4 - x)$ means 'take away 3 times everything inside the bracket': remember that $(-3) \times (-x) = +3x$

EXERCISE 1b

Simplify

1. $2x^2 - 4x + x^2$

2. $5a - 4(a + 3)$

3. $2y - y(x - y)$

4. $8pq - 9p^2 - 3pq$

5. $4xy - y(x - y)$

6. $x^3 - 2x^2 + x^2 - 4x + 5x + 7$

7. $t^2 - 4t + 3 - 2t^2 + 5t + 2$

8. $2(a^2 - b) - a(a + b)$

9. $3 - (x - 4)$
   Note that $-(x - 4)$ means $-1(x - 4)$

10. $5x - 2 - (x + 7)$

11. $3x(x + 2) + 4(3x - 5)$

12. $a(b - c) - c(a - b)$

13. $2cT(3 - T) + 5T(c - 11T)$

14. $x^2(x + 7) - 3x^3 + x(x^2 - 7)$

15. $(3y^2 + 4y - 2) - (7y^2 - 20y + 8)$

16. $6RS + 5RF - R(R + S)$

SIMPLIFICATION OF FRACTIONS

The value of a fraction is unaltered when the *numerator* and *denominator* are each multiplied or divided by the same number.

e.g. $\quad \dfrac{4}{10} = \dfrac{2}{5} = \dfrac{6}{15} = \dfrac{14}{35} = \ldots$

and $\quad \dfrac{ax}{ay} = \dfrac{x}{y} = \dfrac{2x}{2y} = \dfrac{x(p + q)}{y(p + q)} = \ldots$

A fraction can be simplified by multiplying or dividing *top and bottom* by a factor that is common to both.

Example 1c

Simplify $\dfrac{24xy}{18x}$

$$\dfrac{^{4}\cancel{24xy}}{_{3}\cancel{18x}} = \dfrac{4y}{3}$$ Top and bottom are divided by 6 and by x.

EXERCISE 1c

Simplify

1. $\dfrac{2a}{6}$

2. $\dfrac{2a}{a^2}$

3. $\dfrac{pq}{pr}$

4. $\dfrac{12a}{18ab}$

5. $\dfrac{3pq}{8q^2r}$

6. $\dfrac{22ab}{11b}$

7. $\dfrac{18ax^2}{3x}$

8. $\dfrac{36xy}{18y}$

9. $\dfrac{72ab^2}{40a^2b}$

10. $25x^2 \div 15x$

11. $12m^2 \div 6m$

12. $25x^2y \div 5x$

MULTIPLICATION AND DIVISION

Fractions are multiplied by taking the product of the numerators and the product of the denominators,

e.g. $\dfrac{x}{a} \times \dfrac{y}{b} = \dfrac{x \times y}{a \times b} = \dfrac{xy}{ab}$

To divide by a fraction, we multiply by the reciprocal of that fraction, for example

$$\dfrac{x}{a} \div \dfrac{y}{b} = \dfrac{x}{a} \times \dfrac{b}{y} = \dfrac{xb}{ay}$$

Note that fractions can be written in several different ways.

For example, we can write $\dfrac{1}{2a}$ as $1/2a$ or as $1 \div 2a$; these all have the same meaning.

Note that $\frac{1}{2}a$ is *not* the same as $\dfrac{1}{2a}$; $\frac{1}{2}a$ can be written as $\dfrac{a}{2}$.

Examples 1d

1. Simplify $\dfrac{2\pi x^2}{7y^2} \div 4\pi x$

$$\frac{2\pi x^2}{7y^2} \div 4\pi x = \frac{\cancel{2}\,\cancel{\pi}\,x^{\cancel{2}}}{7y^2} \times \frac{1}{_2\cancel{4}\,\cancel{\pi}\,\cancel{x}}$$

$$= \frac{x}{14y^2}$$

2. Simplify $\dfrac{3}{x-2} \div \dfrac{x+1}{x-2}$

$$\frac{3}{(x-2)} \div \frac{(x+1)}{(x-2)} = \frac{3}{\cancel{(x-2)}} \times \frac{\cancel{(x-2)}}{(x+1)}$$

$$= \frac{3}{x+1}$$

Notice that we used brackets to enclose two terms. This makes it easier to see what cancels.

EXERCISE 1d

1. Express $7 \div 4x$ as a fraction.

2. Express $\dfrac{3x}{5}$ as a fraction multiplied by x.

3. Say whether each of the following statements is true or false.

(a) $\frac{1}{4}x = 1/4x$ (b) $3 \div 4p = \frac{3}{4}p$ (d) $\dfrac{a}{3} = \frac{1}{3}a$

Simplify

4. $\dfrac{2}{5} \div \dfrac{1}{x}$

5. $\dfrac{x^2}{y} \div \dfrac{y}{x}$

6. $\dfrac{4x}{y} \times \dfrac{x}{6y}$

7. $2st \times \dfrac{3t}{s^2}$

8. $\dfrac{4uv}{3} \div \dfrac{u}{2v}$

9. $\dfrac{4\pi r^2}{3} \div 2\pi r$

10. $x(2-x) \div \dfrac{2-x}{3}$

11. $\dfrac{3x^2}{2y} \Big/ \dfrac{6xy}{9}$

12. $\dfrac{\pi x^3}{3} \div 8\pi x$

13. $\dfrac{1}{a^2+ab} \Big/ \dfrac{2b}{a}$

14. $\left(\dfrac{1}{a}\right)^2 \times \frac{1}{2}a$

15. $\dfrac{a^2}{3} \times \left(\dfrac{a}{3}\right)^2$

16. $\dfrac{x-4}{x+3} \div \dfrac{2(x-4)}{3}$

17. $\dfrac{x^2}{6} \Big/ \left(\dfrac{x}{2}\right)^2$

18. $\dfrac{2r^3}{3} \times \left(\dfrac{1}{rs}\right)^2$

19. $\dfrac{3x^2}{2y} \times \dfrac{y}{y-2}$

20. $\dfrac{ab}{c} \div \dfrac{ac}{b}$

21. $\dfrac{1}{5}a^2 \Big/ \dfrac{3}{10a}$

COEFFICIENTS

We can identify a particular term in an expression by using the letter, or combination of letters, involved, for example

$2x^2$ is 'the term in x^2'

$3xy$ is 'the term in xy'

The number in front of the letters is called the *coefficient*, for example

in the term $2x^2$, 2 is the coefficient of x^2

in the term $3xy$, 3 is the coefficient of xy

If no number is written in front of a term, the coefficient is 1 or -1, depending on the sign of the term.

Consider the expression $x^3 + 5x^2y - y^3$

the coefficient of x^3 is 1

the coefficient of x^2y is 5

the coefficient of y^3 is -1

There is no term in x^2, so the coefficient of x^2 is zero.

EXERCISE 1e

1. Write down the coefficient of x in $x^2 - 7x + 4$

2. What is the coefficient of xy^2 in the expression $y^3 + 2xy^2 - 7xy$?

3. For the expression $x^2 - 5xy - y^2$ write down the coefficient of
   (a) x^2　　　　　(b) xy　　　　　(c) y^2

4. For the expression $x^3 - 3x + 7$ write down the coefficient of
   (a) x^3　　　　　(b) x^2　　　　　(c) x

EXPANSION OF TWO BRACKETS

Expanding an expression means multiplying it out.

To expand $(2x + 4)(x - 3)$ each term in the first bracket is multiplied by each term in the second bracket. To make sure that nothing is missed out, it is sensible to follow the same order every time. The order used in this book is:

$$(2x + 4)(x - 3) = 2x^2 - 6x + 4x - 12$$
$$= 2x^2 - 2x - 12$$

Use the next exercise to practice expanding and to develop the confidence to go straight to the simplified form.

EXERCISE 1f

Expand and simplify

1. $(x+2)(x+4)$
2. $(x+5)(x+3)$
3. $(a+6)(a+7)$
4. $(t+8)(t+7)$
5. $(s+6)(s+11)$
6. $(2x+1)(x+5)$
7. $(5y+3)(y+5)$
8. $(2a+3)(3a+4)$
9. $(7t+6)(5t+8)$
10. $(11s+3)(9s+2)$
11. $(x-3)(x-2)$
12. $(y-4)(y-1)$
13. $(a-3)(a-8)$
14. $(b-8)(b-9)$
15. $(p-3)(p-12)$
16. $(2y-3)(y-5)$
17. $(x-4)(3x-1)$
18. $(2r-7)(3r-2)$
19. $(4x-3)(5x-1)$
20. $(2a-b)(3a-2b)$
21. $(x-3)(x+2)$
22. $(a-7)(a+8)$
23. $(y+9)(y-7)$
24. $(s-5)(s+6)$
25. $(q-5)(q+13)$
26. $(2t-5)(t+4)$
27. $(x+3)(4x-1)$
28. $(2q+3)(3q-5)$
29. $(x+y)(x-2y)$
30. $(s+2t)(2s-3t)$

Difference of Two Squares

Consider the expansion of $(x-4)(x+4)$,
$$(x-4)(x+4) = x^2 - 4x + 4x - 16$$
$$= x^2 - 16$$

EXERCISE 1g

Expand and simplify

1. $(x-2)(x+2)$
2. $(5-x)(5+x)$
3. $(x+3)(x-3)$
4. $(2x-1)(2x+1)$
5. $(x+8)(x-8)$
6. $(x-a)(x+a)$

From Questions 1 to 6 it is clear that an expansion of the form $(ax+b)(ax-b)$ can be written down directly,

i.e. $(ax+b)(ax-b) = a^2x^2 - b^2$

Use this result to expand the brackets in questions 7 to 12

7. $(x-1)(x+1)$ 8. $(3b+4)(3b-4)$

9. $(2y-3)(2y+3)$ 10. $(ab+6)(ab-6)$

11. $(5x+1)(5x-1)$ 12. $(xy+4)(xy-4)$

Squares

$(2x+3)^2$ means $(2x+3)(2x+3)$

$$\begin{aligned}(2x+3)^2 &= (2x+3)(2x+3)\\&= (2x)^2+(2)(2x)(3)+(3)^2\\&= 4x^2+12x+9\end{aligned}$$

In general

$$\begin{aligned}(ax+b)^2 &= a^2x^2+(2)(ax)(b)+b^2\\&= a^2x^2+2abx+b^2\end{aligned}$$

and $$(ax-b)^2 = a^2x^2-2abx+b^2$$

EXERCISE 1h

Use the results above to expand

1. $(x+4)^2$ 2. $(x+2)^2$ 3. $(2x+1)^2$

4. $(3x+5)^2$ 5. $(2x+7)^2$ 6. $(x-1)^2$

7. $(x-3)^2$ 8. $(2x-1)^2$ 9. $(4x-3)^2$

10. $(5x-2)^2$ 11. $(3t-7)^2$ 12. $(x+y)^2$

13. $(2p+9)^2$ 14. $(3q-11)^2$ 15. $(2x-5y)^2$

Important Expansions

The results from the last two sections should be memorised.   They are summarised here.

$$\begin{aligned}(ax+b)^2 &= a^2x^2+2abx+b^2\\(ax-b)^2 &= a^2x^2-2abx+b^2\\(ax+b)(ax-b) &= a^2x^2-b^2\end{aligned}$$

The next exercise contains a variety of expansions including some of the forms given above.

Example 1i

Expand $(4p + 5)(3 - 2p)$

$$(4p + 5)(3 - 2p) = (5 + 4p)(3 - 2p)$$
$$= 15 + 2p - 8p^2$$

EXERCISE 1i

Expand

1. $(2x - 3)(4 - x)$ **2.** $(x - 7)(x + 7)$

3. $(6 - x)(1 - 4x)$ **4.** $(7p + 2)(2p - 1)$

5. $(3p - 1)^2$ **6.** $(5t + 2)(3t - 1)$

7. $(4 - p)^2$ **8.** $(4t - 1)(3 - 2t)$

9. $(x + 2y)^2$ **10.** $(4x - 3)(4x + 3)$

11. $(3x + 7)^2$ **12.** $(R + 3)(5 - 2R)$

13. $(a - 3b)^2$ **14.** $(2x - 5)^2$

15. $(7a + 2b)(7a - 2b)$ **16.** $(3a + 5b)^2$

17. Write down the coefficients of x^2 and x in the expansion of

    (a) $(2x - 4)(3x - 5)$ (b) $(5x + 2)(3x + 5)$

    (c) $(2x - 3)(7x - 5)$ (d) $(9x + 1)^2$

FACTORISING QUADRATIC EXPRESSIONS

In the last four exercises, each bracket contained a *linear expression*, i.e. an expression that contained an x term and a number term.

An expression of the form $ax + b$, where a and b are numbers, is called a linear expression in x

When two linear expressions in x are multiplied, the result usually contains three terms: a term in x^2, a term in x and a number.

Expressions of this form, i.e. $ax^2 + bx + c$ where a, b and c are numbers and $a \neq 0$, are called *quadratic expressions in x*

Since the product of two linear brackets is quadratic, we might expect to be able to reverse this process. For instance, given a quadratic such as $x^2 - 5x + 6$, we could try to find two linear expressions in x whose product is $x^2 - 5x + 6$. To be able to do this we need to appreciate the relationship between what is inside the brackets and the resulting quadratic.

Consider the examples

$$(2x + 1)(x + 5) = 2x^2 + 11x + 5 \qquad\qquad\qquad [1]$$
$$(3x - 2)(x - 4) = 3x^2 - 14x + 8 \qquad\qquad\qquad [2]$$
$$(x - 5)(4x + 2) = 4x^2 - 18x - 10 \qquad\qquad\qquad [3]$$

The first thing to notice about the quadratic in each example is that

the coefficient of x^2 is the product of the coefficients of x in the two brackets,

the number is the product of the numbers in the two brackets,

the coefficient of x is the sum of the coefficients formed by multiplying the x term in one bracket by the number term in the other bracket.

The next thing to notice is the relationship between the signs.

Positive signs throughout the quadratic come from positive signs in both brackets, as in [1].

A positive number term and a negative coefficient of x in the quadratic come from a negative sign in each bracket, as in [2].

A negative number term in the quadratic comes from a negative sign in one bracket and a positive sign in the other, as in [3].

Examples 1j

1. Factorise $x^2 - 5x + 6$

    The x term in each bracket is x as x^2 can only be $x \times x$
    The sign in each bracket is $-$, so $x^2 - 5x + 6 = (x - \quad)(x - \quad)$
    The numbers in the brackets could be   6 and 1   or   2 and 3
    Checking the middle term tells us that the numbers must be 2 and 3

    $$x^2 - 5x + 6 = (x - 2)(x - 3)$$

    Mentally expanding the brackets checks that they are correct.

2. Factorise $x^2 - 3x - 10$

    The x term in each bracket is x $\Rightarrow$ $x^2 - 3x - 10 = (x - \quad)(x + \quad)$
    The numbers could be   10 and 1   or   5 and 2
    Checking the middle term shows that they are 5 and 2

    $$x^2 - 3x - 10 = (x - 5)(x + 2)$$

    Mentally expanding the brackets confirms that they are correct.

Factorise

1. $x^2 + 8x + 15$
2. $x^2 + 11x + 28$
3. $x^2 + 7x + 6$

4. $x^2 + 7x + 12$
5. $x^2 - 10x + 9$
6. $x^2 - 6x + 9$

7. $x^2 + 8x + 12$
8. $x^2 - 9x + 8$
9. $x^2 + 5x - 14$

10. $x^2 + x - 12$
11. $x^2 - 4x - 5$
12. $x^2 - 10x - 24$

13. $x^2 + 9x + 14$
14. $x^2 - 2x + 1$
15. $x^2 - 9$

16. $x^2 + 5x - 24$
17. $x^2 + 4x + 4$
18. $x^2 - 1$

19. $x^2 - 3x - 18$
20. $x^2 + 10x + 25$
21. $x^2 - 16$

22. $4 + 5x + x^2$
23. $2x^2 - 3x + 1$
24. $3x^2 + 4x + 1$

25. $9x^2 - 6x + 1$
26. $6x^2 - x - 1$
27. $9 + 6x + x^2$

28. $4x^2 - 9$
29. $x^2 + 2ax + a^2$
30. $x^2y^2 - 2xy + 1$

Harder Factorising

When the number of possible combinations of terms for the brackets increases, common sense considerations can help to reduce the possibilities.

For example, if the coefficient of x in the quadratic is odd, then there must be an even number and an odd number in the brackets.

Example 1k

Factorise $12 - x - 6x^2$

The x terms in the brackets could be $6x$ and x, or $3x$ and $2x$, one positive and the other negative.

The number terms could be 12 and 1 or 3 and 4 (not 6 and 2 because the coefficient of x in the quadratic is odd).

Now we try various combinations until we find the correct one.

$$12 - x - 6x^2 = (3 + 2x)(4 - 3x)$$

EXERCISE 1k

Factorise

1. $6x^2 + x - 12$

2. $4x^2 - 11x + 6$

3. $4x^2 + 3x - 1$

4. $3x^2 - 17x + 10$

5. $4x^2 - 12x + 9$

6. $3 - 5x - 2x^2$

7. $25x^2 - 16$

8. $3 - 2x - x^2$

9. $5x^2 - 61x + 12$

10. $9x^2 + 30x + 25$

11. $3 + 2x - x^2$

12. $12 + 7x - 12x^2$

13. $1 - x^2$

14. $9x^2 + 12x + 4$

15. $x^2 + 2xy + y^2$

16. $1 - 4x^2$

17. $4x^2 - 4xy + y^2$

18. $9 - 4x^2$

19. $36 + 12x + x^2$

20. $40x^2 - 17x - 12$

21. $7x^2 - 5x - 150$

22. $36 - 25x^2$

23. $x^2 - y^2$

24. $81x^2 - 36xy + 4y^2$

25. $49 - 84x + 36x^2$

26. $25x^2 - 4y^2$

27. $36x^2 + 60xy + 25y^2$

28. $4x^2 - 4xy - 3y^2$

29. $6x^2 + 11xy + 4y^2$

30. $49p^2q^2 - 28pq + 4$

Common Factors

Consider $4x^2 + 8x + 4$

$$4x^2 + 8x + 4 = 4(x^2 + 2x + 1)$$

The quadratic inside the bracket now has smaller coefficients and can be factorised more easily:

$$4x^2 + 8x + 4 = 4(x + 1)(x + 1)$$
$$= 4(x + 1)^2$$

Not all quadratics factorise.

Consider $3x^2 - x + 5$

The options we can try are $(3x - 5)(x - 1)$ [1]

and $(3x - 1)(x - 5)$ [2]

From [1], $(3x - 5)(x - 1) = 3x^2 - 8x + 5$

From [2], $(3x - 1)(x - 5) = 3x^2 - 16x + 5$

As neither of the possible pairs of brackets expand to give $3x^2 - x + 5$, we conclude that $3x^2 - x + 5$ has no factors of the form $ax + b$ where a and b are integers.

Example 1l

Factorise (a) $2x^2 - 8x + 16$ (b) $x^3 - 3x^2 - 4x$

(a) $2x^2 - 8x + 16 = 2(x^2 - 4x + 8)$

The possible brackets are $(x-1)(x-8)$ and $(x-2)(x-4)$
Neither pair expands to $x^2 - 4x + 8$, so there are no further factors.

(b)   Each term has a factor x, i.e. x is a common factor
$$x^3 - 3x^2 - 4x = x(x^2 - 3x - 4)$$
$$= x(x-4)(x+1)$$

EXERCISE 1l

Factorise where possible

1. $x^2 + x + 1$
2. $2x^2 + 4x + 2$
3. $x^2 + 3x + 2$
4. $3x^2 + 12x - 15$
5. $x^2 + 4$
6. $x^2 - 4x - 6$
7. $x^2 + 3x + 1$
8. $2x^2 - 8x + 8$
9. $3x^2 - 3x - 6$
10. $2x^2 - 6x + 8$
11. $3x^2 - 6x - 24$
12. $x^2 - 4x - 12$
13. $x^2 + 1$
14. $4x^2 - 100$
15. $5x^2 - 25$
16. $7x^2 + x + 4$
17. $10x^2 - 39x - 36$
18. $x^2 + xy + y^2$
19. $x^3 - 4x^2 + 3x$
20. $x^3 - 4x$
21. $x^2y + 2xy + 2y$
22. $x^3 + x$
23. $x^2y^2 - xy$
24. $pq^3 - p^3q$

Further Simplification

When simplifying algebraic fractions, it is sensible to factorise numerators and denominators where possible; look both for common factors and for factors of quadratic expressions.

Examples 1m

1. Simplify $\dfrac{2a^2 - 2ab}{6ab - 6b^2}$

$$\frac{2a^2 - 2ab}{6ab - 6b^2} = \frac{\cancel{2}a\cancel{(a-b)}}{3\cancel{6}b\cancel{(a-b)}}$$

$$= \frac{a}{3b}$$

2. Simplify $\dfrac{\frac{1}{4}x^2 - 1}{x + 2}$

$$\frac{\frac{1}{4}x^2 - 1}{x + 2} = \frac{x^2 - 4}{4(x + 2)}$$

$$= \frac{(x - 2)\cancel{(x + 2)}}{4\cancel{(x + 2)}} = \frac{x - 2}{4}$$

3. Simplify $\dfrac{1}{x^2 - y^2} \div \dfrac{1}{x + y}$

$$\frac{1}{x^2 - y^2} \div \frac{1}{x + y} = \frac{1}{x^2 - y^2} \times \frac{x + y}{1}$$

$$= \frac{1}{(x - y)\cancel{(x + y)}} \times \frac{\cancel{(x + y)}}{1} = \frac{1}{x - y}$$

EXERCISE 1m

Simplify where possible

1. $\dfrac{x - 2}{4x - 8}$

2. $\dfrac{2x + 4}{3x - 6}$

3. $\dfrac{2a + 8}{3a + 12}$

4. $\dfrac{3p - 3q}{5p - 5q}$

5. $\dfrac{x^2 + xy}{xy + y^2}$

6. $\dfrac{x - 3p}{2x + p}$

7. $\dfrac{a - 4}{a - 2}$

8. $\dfrac{x^2y + xy^2}{y^2 + \frac{2}{5}xy}$

9. $\dfrac{\frac{1}{3}a - b}{a + \frac{1}{6}b}$

10. $\dfrac{2x(b - 4)}{6x^2(b + 4)}$

11. $\dfrac{(x - 4)(x - 3)}{x^2 - 16}$

12. $\dfrac{4y^2 + 3}{y^2 - 9}$

13. $\dfrac{\frac{1}{3}(x - 3)}{x^2 - 9}$

14. $\dfrac{x^2 - x - 6}{2x^2 - 5x - 3}$

15. $\dfrac{(x - 2)(x + 2)}{x^2 + x - 2}$

16. $\dfrac{\frac{1}{2}(a + 5)}{a^2 - 25}$

17. $\dfrac{3p + 9q}{p^2 + 6pq + 9q^2}$

18. $\dfrac{a^2 + 2a + 4}{a^2 + 7a + 10}$

19. $\dfrac{x^2 + 2x + 1}{3x^2 + 12x + 9}$

20. $\dfrac{4(x-3)^2}{(x+1)(x^2 - 2x - 3)}$

21. $\dfrac{1}{x^2 - 1} \div \dfrac{1}{x - 1}$

22. $(x+1) \times \dfrac{1}{x^2 - 1}$

23. $\dfrac{x^2 + 4x + 3}{5} \times \dfrac{10}{x + 1}$

24. $\dfrac{4x^2 - 9}{(x-1)^2} \div \dfrac{2x + 3}{x(x-1)}$

25. $\dfrac{x^2 - 4}{2} \div (x^3 - 4x^2 + 4x)$

26. $(x^2 - 2x) \div \dfrac{(x^2 - 3x + 2)}{x}$

HARDER EXPANSIONS

Consider the product $(x - 2)(x^2 - x + 5)$

This expansion should be done in a systematic way.

First multiply each term of the quadratic by x, writing down the separate results as they are found. Then multiply each term of the quadratic by -2. Do not attempt to simplify at this stage.

$$(x - 2)(x^2 - x + 5)$$
$$= x^3 - x^2 + 5x - 2x^2 + 2x - 10$$

Now simplify

$$= x^3 - 3x^2 + 7x - 10$$

Example 1n

Expand $(x + 2)(2x - 1)(x + 4)$

First we expand the last two brackets.

$$(x + 2)(2x - 1)(x + 4) = (x + 2)(2x^2 + 7x - 4)$$
$$= 2x^3 + 7x^2 - 4x + 4x^2 + 14x - 8$$
$$= 2x^3 + 11x^2 + 10x - 8$$

EXERCISE 1n

Expand and simplify

1. $(x-2)(x^2+x+1)$

2. $(3x-2)(x^2-x-1)$

3. $(2x-1)(2x^2-3x+5)$

4. $(x-1)(x^2-x-1)$

5. $(2x+3)(x^2-6x-3)$

6. $(x+1)(x+2)(x+3)$

7. $(x+4)(x-1)(x+1)$

8. $(x-2)(x-3)(x+1)$

9. $(x+1)(2x+1)(x+2)$

10. $(x+2)(x+1)^2$

11. $(2x-1)^2(x+2)$

12. $(3x-1)^3$

13. $(4x+3)(x+1)(x-4)$

14. $(x-1)(2x-1)(2x+1)$

15. $(2x+1)(x+2)(3x-1)$

16. $(x+1)^3$

17. $(x-2)(x+2)(x+1)$

18. $(x+3)(2x+3)(x-1)$

19. $(3x-2)(2x+5)(4x-1)$

20. $2(x-7)(2x+3)(x-5)$

21. Expand and simplify $(x-2)^2(3x-4)$. Write down the coefficients of x^2 and x.

22. Find the coefficients of x^3 and x^2 in the expansion of $(x-4)(2x+3)(3x-1)$

23. Expand and simplify $(x+y)^3$

24. Expand and simplify $(x+y)^4$

PASCAL'S TRIANGLE

It is sometimes necessary to expand expressions such as $(x+y)^4$ but the multiplication is tedious when the power is three or more.

We now describe a far quicker way of obtaining such expansions.

Consider the following expansions,

$$(x+y)^1 = x+y$$
$$(x+y)^2 = x^2+2xy+y^2$$
$$(x+y)^3 = x^3+3x^2y+3xy^2+y^3$$
$$(x+y)^4 = x^4+4x^3y+6x^2y^2+4xy^3+y^4$$

The first thing to notice is that the powers of x and y in the terms of each expansion form a pattern. Looking at the expansion of $(x+y)^4$ we see that the first term is x^4 and then the power of x decreases by 1 in each succeeding term while the power of y increases by 1. For all the terms, the sum of the powers of x and y is 4 and the expansion ends with y^4. There is a similar pattern in the other expansions.

Now consider just the coefficients of the terms. Writing these as a triangular array gives

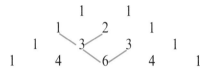

This array is called *Pascal's Triangle* and clearly it has a pattern.

Each row starts and ends with 1 and each other number is the sum of the two numbers in the row above it, as shown. When the pattern is known, Pascal's triangle can be written down to as many rows as needed. Using Pascal's triangle to expand $(x+y)^6$, for example, we go as far as row 6:

$$
\begin{array}{ccccccccccccc}
 & & & & & 1 & & 1 & & & & & \\
 & & & & 1 & & 2 & & 1 & & & & \\
 & & & 1 & & 3 & & 3 & & 1 & & & \\
 & & 1 & & 4 & & 6 & & 4 & & 1 & & \\
 & 1 & & 5 & & 10 & & 10 & & 5 & & 1 & \\
1 & & 6 & & 15 & & 20 & & 15 & & 6 & & 1 \\
\end{array}
$$

We then use our knowledge of the pattern of the powers, together with row 6 of the array, to fill in the coefficients,

i.e. $\quad (x+y)^6 = x^6 + 6x^5y + 15x^4y^2 + 20x^3y^3 + 15x^2y^4 + 6xy^5 + y^6$

The following worked examples show how expansions of other brackets can be found.

Examples 1o

1. Expand $(x+5)^3$

From Pascal's triangle $(x+y)^3 = x^3 + 3x^2y + 3xy^2 + y^3$

Replacing y by 5 gives $(x+5)^3 = x^3 + 3x^2(5) + 3x(5)^2 + (5)^3$

$$= x^3 + 15x^2 + 75x + 125$$

2. Expand $(2x-3)^4$

From Pascal's triangle, $(x+y)^4 = x^4 + 4x^3y + 6x^2y^2 + 4xy^3 + y^4$

Replacing x by $2x$ and y by -3 gives

$$(2x-3)^4 = (2x)^4 + 4(2x)^3(-3) + 6(2x)^2(-3)^2 + 4(2x)(-3)^3 + (-3)^4$$
$$= 16x^4 - 96x^3 + 216x^2 - 216x + 81$$

EXERCISE 1o

Expand

1. $(x+3)^3$

2. $(x-2)^4$

3. $(x+1)^4$

4. $(2x+1)^3$

5. $(x-3)^5$

6. $(p-q)^4$

7. $(2x+3)^3$

8. $(x-4)^5$

9. $(3x-1)^4$

10. $(1+5a)^4$

11. $(2a-b)^6$

12. $(2x-5)^3$

MIXED EXERCISE 1

1. Find the coefficient of x in the expansion of $(3x-7)(5x+4)$

2. Expand $5(3x-2)(3-7x)$

3. Write down the coefficient of y^2 in the expansion of $(2y+9)^3$

4. Factorise $3x^2 - 9x + 6$

5. Write down the coefficient of x^3 in the expansion of $(x-5)^5$

6. Factorise $4x^2 - 36$

7. Expand $x(2x-1)^2$

8. Find the factors of $25 + x^2 - 10x$

9. Find the coefficient of x in the expansion of $(x-4)(3-x)$

10. Expand $3x(2x-1)(2x+1)$

11. Factorise $4x^2 - 12xy + 9y^2$

12. Find the coefficient of xy in the expansion of $(3x-y)(y-4x)$

13. Write down the coefficient of x^2y in the expansion of $(2x-y)^3$

14. Factorise $6x^2 - 11xy - 10y^2$

15. Find the coefficient of a^2b^2 in the expansion of $(2a-b)^4$

16. Simplify (a) $\dfrac{x^2 - 9}{2x - 6}$ (b) $\dfrac{1}{x^2 - 9} \div \dfrac{1}{x + 3}$

17. Simplify $\dfrac{2n - 4}{3} \div (n^2 - 4)$

18. Simplify $\dfrac{x^2 - 9}{x^2 - 5x + 6}$

19. Simplify $(x^2 - 1) \div \left(\dfrac{x + 1}{x - 1}\right)$

20. Simplify $\left(\dfrac{x - 1}{x + 1}\right)^2 \times (x^2 - 1)$

21. Simplify and factorise $(x - 2)(x - 3) - 2$

22. Simplify and factorise $4 - (1 - x)(3 + x)$

23. Factorise $x^2 y - y$

24. Factorise (a) $10 - 3x - x^2$ (b) $x^3 - 4x^2 - 12x$

*25. Various ways of making open-topped cardboard storage boxes are being investigated. One option for making such a box is to start with a square sheet of card, then cut out corners from the sheet, fold along the dotted lines and glue as shown to make the finished box.

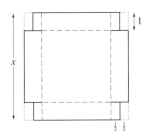 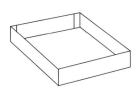

The manufacturers require a comparison of the volume of the box with its weight. This is to be obtained by finding the volume as a fraction of the area of card used to assemble the box. Obtain this expression for the design above by working through the following steps.

(a) Find an expression in terms of x for the area of the base of the box.

(b) Find an expression for the volume of the box.

(c) Find the area of card removed from the sheet before the box is assembled.

(d) Now find an expression for the volume as a fraction of the area of the card used for the assembled box.

CHAPTER 2

ERRORS, SURDS AND INDICES

ERRORS

When we apply mathematical methods to real life situations, it is inevitable that the results we calculate will contain errors. These errors come from many sources and we will look at a few of them here.

Errors are introduced when numerical values are given to quantities such as length and time, for example. First these quantities have to be measured and the measuring instruments used, however sophisticated, are subject to error. Secondly, such quantities cannot be measured exactly; they can only be given to a number of significant figures. Sometimes these values are corrected to a number of significant figures and sometimes they are truncated by reading only the first four figures, say, from a display. When inexact values are used in calculations, further errors are generated. Some quantities, on the other hand, do have exact values; the number of sides in a square for example is exactly four. Working with numerical values yields further numerical values and it is important to know when these are exact and when they contain an error. An error may arise from the nature of the calculation; for example, the result of a calculation with exact numbers could be $\sqrt{2}$ and it is not possible to represent $\sqrt{2}$ exactly as a fraction or a decimal. Alternatively it may arise from recording or storing a fraction such as $\frac{1}{3}$ in decimal form for future use in a calculator or computer. However error arises, it is important to be able to judge the possible size of the error.

Errors can, and almost certainly will, be introduced when numerical values are recorded. It is unrealistic to expect any person not to make any mistakes when reading measuring instruments, writing down numbers or entering numbers into, say, a computer program.

Rounding and Truncation Errors

Suppose that the length of a leaf is measured as 26 mm correct to the nearest millimetre.

Assuming that the measuring instrument is reliable to this degree of accuracy, then the actual length, l mm, of the leaf can be any length from 25.5 mm up to 26.5 mm,

i.e. $25.5 \leqslant l < 26.5$

The number at the lower end of a range is called the *lower bound* and the number at the top end is called the *upper bound*. So the lower bound of this length is 25.5 mm and the upper bound is 26.5 mm.

The error, E, is defined as

given value − true value

Therefore the error, E mm, in giving the length as 26 mm is $(26 - l)$ mm.

Now the greatest value that l can have is just less than 26.5 and in this case E has its lowest value which is just over −0.5.

On the other hand, if l has its lowest value, then E has its greatest value of 0.5,

i.e. $-0.5 \leqslant E \leqslant 0.5$

Absolute Error

The *difference* between the value used and the true value is called the *absolute error*. In the case of the leaf the length is given as 26 mm, and this differs from the true value by at most 0.5 mm. So the maximum absolute error is 0.5 mm.

Relative Error

The relative error is defined as $\dfrac{\text{error}}{\text{true value}}$.

In the case of the leaf we do not know the true length and hence we do not know the error, E. We do know however that $-0.5 \leqslant E \leqslant 0.5$. So the maximum possible value of E is 0.5 and this occurs when l has its lowest possible value of 25.5.

Therefore, in giving the length as 26 mm to the nearest millimetre, the greatest possible value for the relative error is $\dfrac{0.5}{25.5} = -0.0196\ldots$

Similarly the lowest possible relative error is $\dfrac{-0.5}{26.5} = -0.1886\ldots$

Therefore the relative error lies between $-0.018\,86\ldots$ and $0.019\,60\ldots$ Relative error is often given as a percentage. The percentage error in this case lies between -1.9% and 2.0% correct to 2 sf.

Note that relative error is sometimes defined as absolute error over true value.

Calculating with Rounded Numbers

Suppose that the length, l cm, of a rectangle is given as 26 cm and the width, w cm, as 15 cm, both correct to the nearest cm.

$$\therefore \qquad 25.5 \leqslant l < 26.5 \quad \text{and} \quad 14.5 \leqslant w < 15.5$$

The absolute error for both the length and the width is less than or equal to 0.5 cm.

The relative error in the length lies between $\dfrac{-0.5}{26.5}$ and $\dfrac{0.5}{25.5}$,

i.e. between -1.9% and 2% to 2 sf.

The relative error in the width lies between $\dfrac{-0.5}{15.5}$ and $\dfrac{0.5}{14.5}$,

i.e. between -3.2% and 3.4%

(We would expect the relative error in the width to be bigger because 0.5 cm is a larger proportion of 15 cm than it is of 26 cm.)

The area of the rectangle is lw, where $25.5 \times 14.5 \leqslant lw < 26.5 \times 15.5$

$$\Rightarrow \qquad 369.75 \leqslant lw < 410.75$$

Using the given figures, the area is 390 cm^2

The error, E, is such that $-20.75 < E \leqslant 20.25$

The maximum absolute error is 20.75 cm^2.

The percentage error lies between $\dfrac{-20.75}{410.75} \times 100\%$ and $\dfrac{20.25}{369.75} \times 100\%$

i.e. between -5.1% and 5.5%

This shows that multiplying corrected numbers *increases* error.

Now suppose that the length and width were not corrected but were rounded *down* (i.e. truncated) to the nearest complete centimetre, e.g., 26.8 cm is rounded down 26 cm. In this case,

$$26 \leqslant l < 27 \quad \text{and} \quad 15 \leqslant w < 16 \qquad \Rightarrow \qquad 390 \leqslant lw < 432$$

The maximum absolute error in the area is 42 cm^2 and the relative error is between 0 and $-42/390$, i.e. between 0 and -10.8%

We see that using truncated numbers is likely to give a bigger error when they are multiplied than is the case when using corrected numbers.

EXERCISE 2a

1. Two sides of a field need fencing. One side is 45 m long and the other side is 63 m long. Find the maximum absolute error when these values are used to calculate the length of fencing needed if

    (a) the lengths are given to the nearest metre

    (b) the lengths are given rounded down to the nearest complete metre

    (c) the lengths are given rounded up to the next complete metre.

2. The lengths of two pins are 12 mm and 14 mm respectively. Find the maximum absolute error in

    (i) each length   (ii) the difference in their lengths, if

    (a) the lengths are given correct to the nearest mm

    (b) the lengths are given rounded down to the nearest complete mm.

3. Two values, x and y, are given as 1.2 and 3.5 respectively, correct to 2 significant figures. Find

    (a) the upper and lower bounds of

    (i) x (ii) y (iii) xy (iv) $\dfrac{x}{y}$

    (b) the relative percentage error in using the given values to calculate

    (i) xy (ii) x/y

4. Repeat questions 3 when $x = 1.6$ and $y = 0.5$, both numbers having been truncated to 1 decimal place.

5. The calculation $\dfrac{27}{\sqrt{49 + 64}}$ can be done on a calculator either in one stage, using brackets or memory, or in two stages as follows.

    First find $\sqrt{49 + 64}$, write down the answer correct to 3 sf then clear the calculator and find $27 \div$ recorded answer to the first stage.

    (a) Find the percentage error introduced when the second method is used.

    (b) Using the second method, how many significant figures should be recorded for the intermediate stage to give an answer that is correct (rather than corrected) to 3 sf?

6. A triangular piece of land is surveyed. In the triangle, lettered ABC, $\widehat{A}$ is measured as 90°, AC is measured as 250 m to the nearest metre and $\widehat{C}$ is measured as 65° to the nearest degree. Find the greatest absolute error possible when using these measurements to calculate the length of AB.

SQUARE ROOTS

When we express a number as the product of two equal factors, that factor is called the *square root* of the number, for example

$$4 = 2 \times 2 \Rightarrow 2 \text{ is the square root of } 4$$

This is written $\qquad 2 = \sqrt{4}$

Now 4 is also equal to -2×-2 so -2 is also a square root of 4.

The symbol $\sqrt{}$ is used *only for the positive square root,* so we do *not* write $\sqrt{4}$ as -2.

So, although $x^2 = 4 \Rightarrow x = \pm 2$, the only value of $\sqrt{4}$ is 2

The negative square root of 4 is written as $-\sqrt{4}$, and when both square roots are wanted we write $\pm\sqrt{4}$

CUBE ROOTS

When a number can be expressed as the product of three equal factors, that factor is called the *cube root* of the number,

e.g. $\qquad 27 = 3 \times 3 \times 3 \qquad$ so 3 is the cube root of 27

This is written $\qquad \sqrt[3]{27} = 3$

OTHER ROOTS

The notation used for square and cube roots can be extended to represent fourth roots, fifth roots, etc,

e.g. $\qquad 16 = 2 \times 2 \times 2 \times 2 \Rightarrow \sqrt[4]{16} = 2$

and $\qquad 243 = 3 \times 3 \times 3 \times 3 \times 3 \Rightarrow \sqrt[5]{243} = 3$

In general, if a number, n, can be expressed as the product of p equal factors then each factor is called the pth root of n and is written $\sqrt[p]{n}$

RATIONAL NUMBERS

A number which is either an integer, or a fraction whose numerator and denominator are both integers, is called a *rational number*.

The square roots of certain numbers are rational,

e.g. $\quad \sqrt{9} = 3, \quad \sqrt{25} = 5, \quad \sqrt{\frac{4}{49}} = \frac{2}{7}$

This is not true of all square roots however, e.g. $\sqrt{2}$, $\sqrt{5}$, $\sqrt{11}$ are not rational numbers. Such square roots can be given to as many decimal places as are required, for example

$$\sqrt{3} = 1.73 \qquad \text{correct to } 2\,\text{dp}$$

$$\sqrt{3} = 1.732\,05 \quad \text{correct to } 5\,\text{dp}$$

but they can never be expressed exactly as a decimal. They are called *irrational numbers*.

In the first section of this chapter we saw that using corrected numbers in further calculations tends to increase the error in the final result. It is clearly important that such error generation is kept to a minimum and in order to achieve this, numerical results should be left in an exact form whenever this is possible. When irrational numbers are involved, the only way to give them exactly is to leave them in the form $\sqrt{2}$, $\sqrt{7}$ etc; in this form they are called *surds*.

Surds arise in many contexts and, in order to use them for further calculations, you will find it necessary to able to manipulate them.

Simplifying Surds

Consider $\sqrt{18}$

One of the factors of 18 is 9, and 9 has an exact square root,

i.e. $\quad \sqrt{18} = \sqrt{(9 \times 2)} = \sqrt{9} \times \sqrt{2}$

But $\sqrt{9} = 3$, therefore $\sqrt{18} = 3\sqrt{2}$

$3\sqrt{2}$ is the simplest possible surd form for $\sqrt{18}$

Similarly $\quad \sqrt{\dfrac{2}{25}} = \dfrac{\sqrt{2}}{\sqrt{25}} = \dfrac{\sqrt{2}}{5}$

EXERCISE 2b

Express in terms of the simplest possible surd

1. $\sqrt{12}$ 2. $\sqrt{32}$ 3. $\sqrt{27}$ 4. $\sqrt{50}$

5. $\sqrt{200}$ 6. $\sqrt{72}$ 7. $\sqrt{162}$ 8. $\sqrt{288}$

9. $\sqrt{75}$ 10. $\sqrt{48}$ 11. $\sqrt{500}$ 12. $\sqrt{20}$

Multiplying Surds

Consider $(4 - \sqrt{5})(3 + \sqrt{2})$

The multiplication is carried out in the same way and order as when multiplying two linear brackets,

i.e. $(4 - \sqrt{5})(3 + \sqrt{2}) = (4)(3) + (4)(\sqrt{2}) - (3)(\sqrt{5}) - (\sqrt{5})(\sqrt{2})$
$$= 12 + 4\sqrt{2} - 3\sqrt{5} - \sqrt{5}\sqrt{2}$$
$$= 12 + 4\sqrt{2} - 3\sqrt{5} - \sqrt{10}$$

In this example there are no like terms to collect but if the same surd occurs in each bracket the expansion can be simplified.

Examples 2c

1. **Expand and simplify** $(2 + 2\sqrt{7})(5 - \sqrt{7})$

$$(2 + 2\sqrt{7})(5 - \sqrt{7}) = (2)(5) - (2)(\sqrt{7}) + (5)(2\sqrt{7}) - (2\sqrt{7})(\sqrt{7})$$
$$= 10 - 2\sqrt{7} + 10\sqrt{7} - 14$$
$$= 8\sqrt{7} - 4$$

2. **Expand and simplify** $(4 - \sqrt{3})(4 + \sqrt{3})$

$$(4 - \sqrt{3})(4 + \sqrt{3}) = 16 + 4\sqrt{3} - 4\sqrt{3} - (\sqrt{3})(\sqrt{3})$$
$$= 16 - 3$$
$$= 13$$

Example 2 above is a special case because the result is a single rational number. The reader will notice that the two given brackets were of the form $(x - a)(x + a)$, i.e. the factors of $a^2 - x^2$.

Similarly the product of any two brackets of the type $(p - \sqrt{q})(p + \sqrt{q})$ is $p^2 - (\sqrt{q})^2 = p^2 - q$, which is always rational.

This property has an important application in a later section of this chapter.

EXERCISE 2c

Expand and simplify where this is possible.

1. $\sqrt{3}(2 - \sqrt{3})$

2. $\sqrt{2}(5 + 4\sqrt{2})$

3. $\sqrt{5}(2 + \sqrt{75})$

4. $\sqrt{2}(\sqrt{32} - \sqrt{8})$

5. $(\sqrt{3} + 1)(\sqrt{2} - 1)$

6. $(\sqrt{3} + 2)(\sqrt{3} + 5)$

7. $(\sqrt{5} - 1)(\sqrt{5} + 1)$

8. $(2\sqrt{2} - 1)(\sqrt{2} - 1)$

9. $(\sqrt{5} - 3)(2\sqrt{5} - 4)$

10. $(4 + \sqrt{7})(4 - \sqrt{7})$

11. $(\sqrt{6} - 2)^2$

12. $(2 + 3\sqrt{3})^2$

Multiply by a bracket which will make the product rational.

13. $(1 + \sqrt{2})$

14. $(1 - \sqrt{3})$

15. $(\sqrt{2} - 2)$

16. $(\sqrt{5} + 1)$

17. $(4 - \sqrt{5})$

18. $(\sqrt{11} + 3)$

19. $(2\sqrt{3} - 4)$

20. $(\sqrt{6} - \sqrt{5})$

21. $(3 - 2\sqrt{3})$

22. $(2\sqrt{5} - \sqrt{2})$

Rationalising a Denominator

A fraction whose denominator contains a surd is more awkward to deal with than one where a surd occurs only in the numerator.

There is a technique for transferring the surd expression from the denominator to the numerator; it is called *rationalising the denominator* (i.e. making the denominator into a rational number).

Examples 2d

1. Rationalise the denominator of $\dfrac{2}{\sqrt{3}}$

The square root in the denominator can be removed if we multiply it by another $\sqrt{3}$. If this is done we must, of course, multiply the numerator also by $\sqrt{3}$, otherwise the value of the fraction is changed.

$$\frac{2}{\sqrt{3}} = \frac{2\sqrt{3}}{(\sqrt{3})(\sqrt{3})} = \frac{2\sqrt{3}}{3}$$

2. Rationalise the denominator and simplify $\dfrac{3\sqrt{2}}{5-\sqrt{2}}$

We saw in Examples 2b number 2, that a product of the type $(a-\sqrt{b})(a+\sqrt{b})$ is wholly rational, so in this question we multiply numerator and denominator by $5+\sqrt{2}$

$$\frac{3\sqrt{2}}{5-\sqrt{2}} = \frac{3\sqrt{2}(5+\sqrt{2})}{(5-\sqrt{2})(5+\sqrt{2})}$$

$$= \frac{15\sqrt{2}+3(\sqrt{2})(\sqrt{2})}{25-(\sqrt{2})(\sqrt{2})}$$

$$= \frac{15\sqrt{2}+6}{23}$$

EXERCISE 2d

Rationalise the denominator, simplifying where possible.

1. $\dfrac{3}{\sqrt{2}}$

2. $\dfrac{1}{\sqrt{7}}$

3. $\dfrac{2}{\sqrt{11}}$

4. $\dfrac{3\sqrt{2}}{\sqrt{5}}$

5. $\dfrac{1}{\sqrt{27}}$

6. $\dfrac{\sqrt{5}}{\sqrt{10}}$

7. $\dfrac{1}{\sqrt{2}-1}$

8. $\dfrac{3\sqrt{2}}{5+\sqrt{2}}$

9. $\dfrac{2}{2\sqrt{3}-3}$

10. $\dfrac{5}{2-\sqrt{5}}$

11. $\dfrac{1}{\sqrt{7}-\sqrt{3}}$

12. $\dfrac{4\sqrt{3}}{2\sqrt{3}-3}$

13. $\dfrac{3-\sqrt{5}}{\sqrt{5}+1}$

14. $\dfrac{2\sqrt{3}-1}{4-\sqrt{3}}$

15. $\dfrac{\sqrt{5}-1}{\sqrt{5}-2}$

16. $\dfrac{3}{\sqrt{3}-\sqrt{2}}$

17. $\dfrac{3\sqrt{5}}{2\sqrt{5}+1}$

18. $\dfrac{\sqrt{2}+1}{\sqrt{2}-1}$

19. $\dfrac{2\sqrt{7}}{\sqrt{7}+2}$

20. $\dfrac{\sqrt{5}-1}{3-\sqrt{5}}$

21. $\dfrac{1}{\sqrt{11}-\sqrt{7}}$

22. $\dfrac{4-\sqrt{3}}{3-\sqrt{3}}$

23. $\dfrac{1-3\sqrt{2}}{3\sqrt{2}+2}$

24. $\dfrac{1}{3\sqrt{2}-2\sqrt{3}}$

25. $\dfrac{\sqrt{3}}{\sqrt{2}(\sqrt{6}-\sqrt{3})}$

26. $\dfrac{1}{\sqrt{3}(\sqrt{21}+\sqrt{7})}$

27. $\dfrac{\sqrt{2}}{\sqrt{3}(\sqrt{5}-\sqrt{2})}$

INDICES

Base and Index

In an expression such as 3^4, the *base* is 3 and the 4 is called the *power* or *index* (the plural is *indices*).

Working with indices involves using some properties which apply to any base, so we express these rules in terms of a general base a (i.e. a stands for any number).

Rule 1
Because a^3 means $a \times a \times a$ and a^2 means $a \times a$ it follows that

$$a^3 \times a^2 = (a \times a \times a) \times (a \times a) = a^5$$

i.e. $\quad a^3 \times a^2 = a^{3+2}$

Similar examples with different powers all indicate the general rule that

$$\boldsymbol{a^p \times a^q = a^{p+q}}$$

Rule 2
Now dealing with division we have

$$a^7 \div a^4 = \frac{\cancel{a} \times \cancel{a} \times \cancel{a} \times \cancel{a} \times a \times a \times a}{\cancel{a} \times \cancel{a} \times \cancel{a} \times \cancel{a}} = a^3$$

i.e. $\quad a^7 \div a^4 = a^{7-4}$

Again this is just one example of the general rule

$$\boldsymbol{a^p \div a^q = a^{p-q}}$$

When this rule is applied to certain fractions some interesting cases arise.

Consider $a^3 \div a^5$

$$\frac{a^3}{a^5} = \frac{\cancel{a} \times \cancel{a} \times \cancel{a}}{\cancel{a} \times \cancel{a} \times \cancel{a} \times a \times a} = \frac{1}{a^2}$$

But from Rule 2 we have

$$a^3 \div a^5 = a^{3-5} = a^{-2}$$

Therefore a^{-2} means $\dfrac{1}{a^2}$

In general
$$a^{-p} = \frac{1}{a^p}$$

i.e. $\qquad\qquad a^{-p}$ means 'the reciprocal of a^p'

Now consider $a^4 \div a^4$

$$\frac{a^4}{a^4} = \frac{\cancel{a} \times \cancel{a} \times \cancel{a} \times \cancel{a}}{\cancel{a} \times \cancel{a} \times \cancel{a} \times \cancel{a}} = 1$$

From Rule 2, $\qquad \dfrac{a^4}{a^4} = a^{4-4} = a^0$

Therefore $\qquad\qquad\qquad a^0 = 1$

i.e. $\qquad\qquad$ **any base to the power zero is equal to 1**

Rule 3

$$(a^2)^3 = (a \times a)^3$$
$$= (a \times a) \times (a \times a) \times (a \times a)$$
$$= a^6$$

i.e. $\quad (a^2)^3 = a^{2 \times 3}$

In general $\qquad\qquad (a^p)^q = a^{pq}$

Rule 4
This rule explains the meaning of a fractional index.
From the first rule we have
$$a^{1/2} \times a^{1/2} = a^{1/2+1/2} = a^1 = a$$
i.e. $\qquad\qquad a = a^{1/2} \times a^{1/2}$
But $\qquad\qquad a = \sqrt{a} \times \sqrt{a}$
Therefore $a^{1/2}$ means $\sqrt{a}$, i.e. the positive square root of a
Similarly $a^{1/3} \times a^{1/3} \times a^{1/3} = a^{1/3+1/3+1/3} = a^1 = a$
But $\qquad \sqrt[3]{a} \times \sqrt[3]{a} \times \sqrt[3]{a} = a$
Therefore $a^{1/3}$ means $\sqrt[3]{a}$, i.e. the cube root of a

In general $\qquad\qquad a^{1/p} = \sqrt[p]{a}$, **i.e. the p th root of a**

For a more general fractional index, $\frac{p}{q}$, the third rule shows that

$$a^{p/q} = (a^p)^{1/q} \quad \text{or} \quad (a^{1/q})^p$$

For example

$$a^{3/4} = (a^3)^{1/4} \quad \text{or} \quad (a^{1/4})^3$$
$$= \sqrt[4]{a^3} \quad \text{or} \quad (\sqrt[4]{a})^3$$

i.e. $a^{3/4}$ represents either 'the fourth root of a^3'
or 'the cube of the fourth root of a'

All the general rules can be applied to simplify a wide range of expressions containing indices *provided that the terms all have the same base.*

Examples 2e

1. Simplify (a) $\dfrac{2^3 \times 2^7}{4^3}$ (b) $(x^2)^7 \times x^{-3}$ (c) $\sqrt[3]{(a^4 b^5)} \times b^{1/3}/a$

(a) First we express all the terms to a base 2

$$\frac{2^3 \times 2^7}{4^3} = \frac{2^3 \times 2^7}{(2^2)^3}$$
$$= \frac{2^{3+7}}{2^{2 \times 3}}$$
$$= \frac{2^{10}}{2^6} = 2^4$$

(b) $$(x^2)^7 \times x^{-3} = x^{2 \times 7} \times \frac{1}{x^3}$$
$$= x^{14} \times \frac{1}{x^3} = x^{11}$$

(c) $$\sqrt[3]{(a^4 b^5)} \times b^{1/3}/a = (a^{4/3})(b^{5/3})(b^{1/3})(a^{-1})$$
$$= (a^{4/3-1})(b^{5/3+1/3})$$
$$= a^{1/3} b^2$$

2. Evaluate (a) $(64)^{-1/3}$ (b) $\left(\dfrac{25}{9}\right)^{-3/2}$

(a) $(64)^{-1/3} = \dfrac{1}{(64)^{1/3}} = \dfrac{1}{\sqrt[3]{64}} = \dfrac{1}{4}$

(b) $\left(\dfrac{25}{9}\right)^{-3/2} = \left(\dfrac{9}{25}\right)^{3/2} = \left(\sqrt{\dfrac{9}{25}}\right)^3 = \left(\dfrac{3}{5}\right)^3 = \dfrac{27}{125}$

Note that $\left(\dfrac{9}{25}\right)^{3/2}$ could have been expressed as $\sqrt{\left(\dfrac{9}{25}\right)^3}$ but this form involves *much* bigger numbers.

EXERCISE 2e
Simplify

1. $(2^3)^2$

2. $(x^4)^2$

3. $(ab^2)^3$

4. $a^3 \div a^5$

5. $2x^2 \times 3x^3$

6. $4(3^3) \times 2(3^2)$

7. $(2^{-1})^2$

8. $(-4)^{-2}$

9. $(x^{-2})^3$

10. $\dfrac{2^4}{2^2 \times 4^3}$

11. $4^{1/2} \times 2^{-3}$

12. $(3^3)^{1/2} \times 9^{1/4}$

13. $\dfrac{x^{1/3} \times x^{4/3}}{x^{-1/3}}$

14. $\dfrac{p^{1/2} \times p^{-3/4}}{p^{-1/4}}$

15. $(\sqrt{t})^3 \times (\sqrt{t^5})$

16. $(y^2)^{3/2} \times y^{-3}$

17. $(16)^{5/4} \div 8^{4/3}$

18. $\dfrac{y^{1/2}}{y^{-3/4}} \times \sqrt{(y^{1/2})}$

19. $x^2 \times x^{5/2} \div x^{-1/2}$

20. $\dfrac{y^{1/6} \times y^{-2/3}}{y^{1/4}}$

21. $(p^{1/3})^2 \times (p^2)^{1/3} \div \sqrt[3]{p}$

Evaluate

22. $(16)^{-1/4}$

23. $(27)^{2/3}$

24. $(81)^{3/4}$

25. $\left(\dfrac{1}{2}\right)^{-3}$

26. $\left(\dfrac{1}{3}\right)^{-2}$

27. $\left(\dfrac{3}{4}\right)^{-2}$

28. $\dfrac{1}{2^{-2}}$

29. $\left(\dfrac{2}{5}\right)^{-3}$

30. $\dfrac{2^{-2}}{2^{-1}}$

31. $\left(\dfrac{1}{3}\right)^{-1}$

32. $\left(\dfrac{1}{4}\right)^{5/2}$

33. $(8)^{-1/3}$

34. $\dfrac{1}{(16)^{-1/4}}$

35. $\left(\dfrac{1}{9}\right)^{-3/2}$

36. $\left(\dfrac{27}{8}\right)^{2/3}$

37. $\left(\dfrac{100}{9}\right)^{0}$

38. $\dfrac{1}{4^{-2}}$

39. $(0.64)^{-1/2}$

40. $\left(-\dfrac{1}{5}\right)^{-1}$

41. $(121)^{3/2}$

42. $\left(\dfrac{125}{27}\right)^{-1/3}$

43. $18^{1/2} \times 2^{1/2}$

44. $3^{-3} \times 2^{0} \times 4^{2}$

45. $\dfrac{8^{1/2} \times 32^{1/2}}{(16)^{1/4}}$

46. $5^{1/3} \times 25^{0} \times 25^{1/3}$

47. $27^{1/4} \times 3^{1/4} \times (\sqrt{3})^{-2}$

48. $\dfrac{9^{1/3} \times 27^{-1/2}}{3^{-1/6} \times 3^{-2/3}}$

MIXED EXERCISE 2

1. Simplify (a) $\sqrt{84}$ (b) $\sqrt{300}$ (c) $\sqrt{45}$
   (d) $\sqrt{27}$ (e) $\sqrt{250}$ (f) $\sqrt{18}$

2. Expand and simplify

   (a) $\sqrt{3}(1 - \sqrt{3})$ (b) $\sqrt{2}(3 - 2\sqrt{2})$
   (c) $\sqrt{3}(7 - 2\sqrt{3})$ (d) $\sqrt{2}(2\sqrt{2} + \sqrt{8})$

3. Expand and simplify (a) $(3 + \sqrt{2})(4 - 2\sqrt{2})$ (b) $(\sqrt{5} - \sqrt{2})^{2}$

4. Multiply by a bracket that will make the product rational
   (a) $(7 - \sqrt{3})$ (b) $(2\sqrt{2} + 1)$ (c) $(\sqrt{7} - \sqrt{5})$

5. Rationalise the denominator and simplify where possible
   (a) $\dfrac{5}{\sqrt{7}}$ (b) $\dfrac{3}{\sqrt{13} - 2}$ (c) $\dfrac{4}{\sqrt{3} - \sqrt{2}}$ (d) $\dfrac{\sqrt{3} - 1}{\sqrt{3} + 1}$

6. Write down the value of
   (a) 2^{5} (b) 3^{3} (c) $243^{1/5}$ (d) 2^{-2} (e) $225^{1/2}$

7. Simplify (a) $\dfrac{2^{3} \times 4^{-2}}{2^{-1}}$ (b) $(x^{3})^{-2} \times (x^{2})^{3}$

8. Evaluate (a) $(64)^{-1/3}$ (b) $\left(\dfrac{49}{16}\right)^{-1/2}$ (c) $\left(\dfrac{8}{27}\right)^{3/2}$

9. Simplify (a) $8^{1/6} \times 2^0 \times 2^{-1/2}$ (b) $(\sqrt{5})^{-2} \times 75^{1/2} \times 25^{-1/4}$

10. What is the value of the missing number or index?

   (a) $9^{3/2} = 3^{\square}$ (b) $2^{\square} = 8^{-1/4}$ (c) $\dfrac{1}{16^{3/2}} = \square^{-3}$

11. The diagram shows a prototype of a wedge.

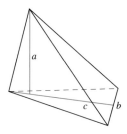

   The values of a, b and c are calculated and found to be given by

   $$a = \left(2 - \frac{2}{\sqrt{3}}\right), \qquad b = (2\sqrt{5} - 4), \qquad c = \frac{3}{\sqrt{3} + 1}$$

   The volume is then found by substituting these values in the formula

   $$V = \tfrac{1}{6} abc$$

   (a) Find, corrected to 3 significant figures, the values of a, b and c.

   (b) Use the values found for a, b and c in (a) to find V, giving the result corrected to 3 significant figures.

   (c) Substitute the exact forms for the values of a, b and c into the formula for V and simplify the result.

   (d) Use the answer to (c) to find the value of V corrected to 3 significant figures.

   (e) Find the percentage error involved in the result from (b) compared with the result from (d).

*12. It can be shown that $\sqrt{2}$ is irrational by using a method known as *proof by contradiction*. This involves making an assumption and then showing that this assumption is incorrect. Here is the argument applied to $\sqrt{2}$.

   Assume that $\sqrt{2}$ is rational.

   i.e. that $\sqrt{2} = \dfrac{a}{b}$ where a and b are integers with no common factors   [1]

   Squaring both sides gives

   $$2 = \frac{a^2}{b^2} \quad \Rightarrow \quad 2b^2 = a^2$$ [2]

Now $2b^2$ must be an even number so [2] shows that a^2, and hence a, is an even number. Therefore a can be replaced by $2c$ where c is an integer.

Substituting $2c$ for a in [2] gives

$$2b^2 = 4c^2 \quad \Rightarrow \quad b^2 = 2c^2$$

This shows that b^2, and hence b, is an even number.

But a and b cannot both be even numbers if [1] is correct, so we deduce that [1] cannot be correct, i.e. $\sqrt{2}$ is not rational.

Use a similar argument to show that $\sqrt{3}$ is also irrational.

CHAPTER 3

LINEAR AND QUADRATIC EQUATIONS AND SIMULTANEOUS EQUATIONS

EXPRESSIONS AND EQUATIONS

Consider the following problem.

Find the number which, when added to itself, gives 28.

This is a trivial problem in the sense that the answer is obvious but, because it is so easy, it is useful for illustrating some basic concepts.

The first step in solving a problem mathematically is to make the unknown items general by using a letter for each of them. In this case only one number is unknown, but it could be *any* number so we represent it by x (or any other letter).

For the next stage the given information is translated into statements containing letters and symbols.

This problem states that x is added to itself: this translates to $x + x$.

Now

any collection of letters and numbers is an algebraic expression.

So $x + x$ is an example of an *algebraic expression.*

We know that x added to itself gives 28.

Therefore $\quad x + x = 28,\quad$ i.e. $\quad 2x = 28$

This is valid only for $x = 14$.

Now

an equality of two expressions that is valid only for certain values is an equation

so $2x = 28$ is an example of an equation.

Now $x = 14$ is called the solution of the equation and the process of finding this particular value is called solving the equation.

In the solution we used the fact that $x + x$ is equivalent to $2x$, and this can be written as $x + x = 2x$.

This equality is true regardless of the value of x, because $2x$ is just another way of writing $x + x$

$x + x = 2x$ is example of an *identity*.

An identity is the equivalence of two forms of the *same* expression.

The symbol '$=$' means 'is equal to' and is always used in an equation. There is another symbol, $\equiv$, which means 'is equivalent to' or 'is identical to', and, strictly speaking, we should use it in identities, e.g. we should write $x + x \equiv 2x$. In practice however, the symbol $=$ is usually used for identities as well as for equations.

Problems that can be translated into mathematical terms usually result in equations that need to be solved. This chapter looks at a variety of equations and algebraic methods of solving them.

There is however one fundamental principle that applies when solving any equation: *the equality must be preserved.* In practice, this means that

if the expression on one side of the equal sign is altered,
the expression on the other side must be altered in exactly the same way.

LINEAR EQUATIONS

A linear equation involves linear expressions only,

e.g. $\quad 2x - 3 = 5, \quad \dfrac{2x - 5}{3} = 7x, \quad \left(\tfrac{1}{3}\right)x + \left(\tfrac{1}{2}\right) - \left(\tfrac{1}{4}\right)(5 - x) = 0.$

Solving a linear equation involves the following processes applied in order as necessary;

- remove all fractions
- expand brackets
- collect terms containing the unknown on one side of the equality (choose the side where the collected terms will have a positive coefficient); collect all the number terms on the other side
- divide by the coefficient in the term containing the unknown.

Example 3a

Solve the equations (a) $\dfrac{2x-5}{x} = 7$ (b) $2\pi x = \pi^2\left(x - \tfrac{1}{2}\right)$

(a) $\dfrac{2x-5}{x} = 7$

Multiplying both sides by x gives	$2x - 5 = 7x$
Subtracting $2x$ from both sides gives	$-5 = 5x$
Dividing both sides by 5 gives	$-1 = x,$ i.e. $x = -1$

Note that in order to multiply by x, the numbers represented by x must exclude zero.

(b) $2\pi x = \pi^2\left(x - \tfrac{1}{2}\right)$

Expanding the bracket gives $\qquad\qquad\qquad 2\pi x = \pi^2 x - \tfrac{1}{2}\pi^2$

Multiplying both sides by 2 gives $\qquad\quad 4\pi x = 2\pi^2 x - \pi^2$

Collecting the terms containing x on the left-hand side and number terms on the right-hand side (π is a number) gives

$$4\pi x - 2\pi^2 x = -\pi^2$$

Factorising the left-hand side gives $\qquad\quad 2x(2\pi - \pi^2) = -\pi^2$

Dividing by the coefficient of x: $\qquad\qquad\quad x = \dfrac{\pi^2}{2(2\pi - \pi^2)}$

The solution can be left in this form but it is worth multiplying top and bottom by -1 to make the numerator positive.

$$= \frac{\pi^2}{2(\pi^2 - 2\pi)}$$

$$= \frac{\pi^2}{2\pi(\pi - 2)} = \frac{\pi}{2(\pi - 2)}$$

Sometimes it is necessary to solve an equation for one unknown in terms of other letters.

For example, to find x when $2x - a = b - ax$, we apply the same procedures as we use when dealing with numbers, i.e.

collect the terms containing x on one side and all other terms on the other side:

$$2x + ax = b + a$$

factorise: $\qquad\qquad\qquad x(2 + a) = b + a$

divide by the coefficient of x: $\qquad x = \dfrac{b + a}{2 + a}$

EXERCISE 3a

Solve the equations for x giving answers in exact form.

1. $2x - 5 = 9$

2. $3x - 4 = 6 - 2x$

3. $3x - x + 5 = 3x + 6 - 3$

4. $2(x - 1) = 3x$

5. $2 - 3(2 - x) = 8$

6. $5(3 - 2x) - 2(1 - x) = 5$

7. $1 - \frac{1}{5}x = 3$

8. $3 \div 2x = 2$

9. $2(x - 3) = \frac{1}{4}$

10. $\frac{x - 1}{x} = 2$

11. $\frac{1}{2}x - \frac{1}{3} = 5(1 - 2x)$

12. $\frac{2}{x} = 0.4$

13. $\frac{3}{x} - \frac{1}{3x} = 2$

14. $\frac{1}{x + 1} + \frac{2}{x} = 0$

15. $\frac{x - 1}{3} = \frac{2x - 3}{4} + x$

16. $ax + b = 4$

17. $\frac{x}{3} + y = \frac{c}{2}$

18. $ax + by + c = 0$

QUADRATIC EQUATIONS

When a quadratic expression has a particular value we have a quadratic equation, for example

$$2x^2 - 5x + 1 = 0$$

Using, a, b and c to stand for any numbers, any quadratic equation can be written in the general form

$$ax^2 + bx + c = 0$$

Solution by Factorising

Consider the quadratic equation $x^2 - 3x + 2 = 0$

The quadratic expression on the left-hand side can be factorised,

i.e. $x^2 - 3x + 2 = (x - 2)(x - 1)$

Therefore the given equation becomes

$$(x - 2)(x - 1) = 0 \tag{1}$$

Now if the product of two quantities is zero then one, or both, of those quantities must be zero.

Applying this fact to equation [1] gives

$$x - 2 = 0 \quad \text{or} \quad x - 1 = 0$$

i.e. $\qquad x = 2 \quad \text{or} \quad x = 1$

This is the solution of the given equation.
The values 2 and 1 are called the *roots* of that equation.

This method of solution can be used for any quadratic equation in which the quadratic expression factorises.

Example 3b

Find the roots of the equation $x^2 + 6x - 7 = 0$

$$x^2 + 6x - 7 = 0$$

$\Rightarrow \qquad (x - 1)(x + 7) = 0$

$\therefore \qquad x - 1 = 0 \quad \text{or} \quad x + 7 = 0$

$\therefore \qquad x = 1 \quad \text{or} \quad x = -7$

The roots of the equation are 1 and -7

EXERCISE 3b

Solve the equations.

1. $x^2 + 5x + 6 = 0$

2. $x^2 + x - 6 = 0$

3. $x^2 - x - 6 = 0$

4. $x^2 + 6x + 8 = 0$

5. $x^2 - 4x + 3 = 0$

6. $x^2 + 2x - 3 = 0$

7. $2x^2 + 3x + 1 = 0$

8. $4x^2 - 9x + 2 = 0$

9. $x^2 + 4x - 5 = 0$

10. $x^2 + x - 72 = 0$

Find the roots of the equations.

11. $x^2 - 2x - 3 = 0$

12. $x^2 + 5x + 4 = 0$

13. $x^2 - 6x + 5 = 0$

14. $x^2 + 3x - 10 = 0$

15. $x^2 - 5x - 14 = 0$

16. $x^2 - 9x + 14 = 0$

Rearranging the Equation

The terms in a quadratic equation are not always given in the order $ax^2 + bx + c = 0$. When they are given in a different order they should be rearranged in the standard form.

For example

$$x^2 - x = 4 \qquad \text{becomes} \qquad x^2 - x - 4 = 0$$
$$3x^2 - 1 = 2x \qquad \text{becomes} \quad 3x^2 - 2x - 1 = 0$$
$$x(x - 1) = 2 \qquad \text{becomes} \quad x^2 - x = 2 \;\Rightarrow\; x^2 - x - 2 = 0$$

It is usually best to collect the terms on the side where the x^2 term is positive, for example

$$2 - x^2 = 5x \quad \text{becomes} \quad 0 = x^2 + 5x - 2$$

i.e. $\qquad x^2 + 5x - 2 = 0$

Losing a Solution

Quadratic equations sometimes have a common factor containing the unknown quantity. It is very tempting in such cases to divide by the common factor, but doing this results in the loss of part of the solution, as the following example shows.

First solution

$$x^2 - 5x = 0$$
$$x(x - 5) = 0$$
$\therefore \qquad x = 0 \quad \text{or} \quad x - 5 = 0$
$\Rightarrow \qquad x = 0 \quad \text{or} \quad 5$

Second solution

$$x^2 - 5x = 0$$
$$x - 5 = 0 \quad (\text{Dividing by } x)$$
$\therefore \qquad x = 5$
The solution $x = 0$ has been lost.

Although dividing an equation by a numerical common factor is correct and sensible,

dividing an equation by a common factor containing the unknown quantity results in the loss of a solution.

Examples 3c

1. Solve the equation $4x - x^2 = 3$

$$4x - x^2 = 3$$
$\Rightarrow \qquad\qquad 0 = x^2 - 4x + 3$
$\Rightarrow \qquad x^2 - 4x + 3 = 0$
$\Rightarrow \qquad (x - 3)(x - 1) = 0$
$\Rightarrow \qquad x - 3 = 0 \quad \text{or} \quad x - 1 = 0$
$\Rightarrow \qquad x = 3 \quad \text{or} \quad x = 1$

2. Find the roots of the equation $x^2 = 3x$

$$x^2 = 3x$$
$$\Rightarrow \quad x^2 - 3x = 0$$
$$\Rightarrow \quad x(x-3) = 0$$
$$\Rightarrow \quad x = 0 \quad \text{or} \quad x - 3 = 0$$
$$\Rightarrow \quad x = 0 \quad \text{or} \quad x = 3$$

Therefore the roots are 0 and 3

EXERCISE 3c

Solve the equations.

1. $x^2 + 10 - 7x = 0$

2. $15 - x^2 - 2x = 0$

3. $x^2 - 3x = 4$

4. $12 - 7x + x^2 = 0$

5. $2x - 1 + 3x^2 = 0$

6. $x(x+7) + 6 = 0$

7. $2x^2 - 4x = 0$

8. $x(4x+5) = -1$

9. $2 - x = 3x^2$

10. $6x^2 + 3x = 0$

11. $x^2 + 6x = 0$

12. $x^2 = 10x$

13. $x(4x+1) = 3x$

14. $20 + x(1-x) = 0$

15. $x(3x-2) = 8$

16. $x^2 - x(2x-1) + 2 = 0$

17. $x(x+1) = 2x$

18. $4 + x^2 = 2(x+2)$

19. $x(x-2) = 3$

20. $1 - x^2 = x(1+x)$

Solution by Completing the Square

When there are no obvious factors, another method is needed to solve the equation. One such method involves adding a constant to the x^2 term and the x term, to make a perfect square. This technique is called *completing the square*.

Consider $\qquad\qquad x^2 - 2x$

Adding 1 gives $\qquad x^2 - 2x + 1$

Now $x^2 - 2x + 1 = (x-1)^2$ which is a perfect square.

Adding the number 1 was not a guess, it was found by using the fact that

$$x^2 + 2ax + \boxed{a^2} = (x+a)^2$$

We see from this that the number to be added is always
(half the coefficient of x)^2

Hence $x^2 + 6x$ requires 3^2 to be added to make a perfect square,

i.e. $\quad x^2 + 6x + 9 = (x+3)^2$

To complete the square when the coefficient of x^2 is not 1, we first take out the coefficient of x^2 as a factor,

e.g. $\quad 2x^2 + x = 2\left(x^2 + \tfrac{1}{2}x\right)$

Now we add $\left(\tfrac{1}{2} \times \tfrac{1}{2}\right)^2$ inside the bracket, giving

$$2\left(x^2 + \tfrac{1}{2}x + \tfrac{1}{16}\right) = 2\left(x + \tfrac{1}{4}\right)$$

Take extra care when the coefficient of x^2 is negative

e.g. $\quad -x^2 + 4x = -(x^2 - 4x)$

Then $\quad -(x^2 - 4x + 4) = -(x-2)^2$

$\therefore \quad -x^2 + 4x - 4 = -(x-2)^2$

Examples 3d

1. Solve the equation $x^2 - 4x - 2 = 0$, giving the solution in surd form.

$$x^2 - 4x - 2 = 0$$

No factors can be found so we isolate the two terms with x in,

i.e. $\quad x^2 - 4x = 2$

Add $\left\{\tfrac{1}{2} \times (-4)\right\}^2$ to *both* sides

i.e. $\quad x^2 - 4x + 4 = 2 + 4$

$\Rightarrow \quad (x-2)^2 = 6$

$\therefore \quad x - 2 = \pm\sqrt{6}$

$\therefore \quad x = 2 + \sqrt{6} \quad \text{or} \quad x = 2 - \sqrt{6}$

2. Find in surd form the roots of the equation $2x^2 - 3x - 3 = 0$

$$2x^2 - 3x - 3 = 0$$
$$2x^2 - 3x = 3$$
$$2\left(x^2 - \tfrac{3}{2}x\right) = 3$$
$$x^2 - \tfrac{3}{2}x = \tfrac{3}{2}$$
$$x^2 - \tfrac{3}{2}x + \tfrac{9}{16} = \tfrac{3}{2} + \tfrac{9}{16}$$
$$\left(x - \tfrac{3}{4}\right)^2 = \tfrac{33}{16}$$
$$\therefore \quad x - \tfrac{3}{4} = \pm\sqrt{\tfrac{33}{16}} = \pm\tfrac{1}{4}\sqrt{33}$$
$$\therefore \quad x = \tfrac{3}{4} \pm \tfrac{1}{4}\sqrt{33}$$

The roots of the equation are $\tfrac{1}{4}(3 + \sqrt{33})$ and $\tfrac{1}{4}(3 - \sqrt{33})$

EXERCISE 3d

Add a number to each expression so that the result contains a perfect square.

1. $x^2 - 4x$ **2.** $x^2 + 2x$ **3.** $x^2 - 6x$

4. $x^2 + 10x$ **5.** $2x^2 - 4x$ **6.** $x^2 + 5x$

7. $3x^2 - 48x$ **8.** $x^2 + 18x$ **9.** $2x^2 - 40x$

10. $x^2 + x$ **11.** $3x^2 - 2x$ **12.** $2x^2 + 3x$

Solve the equations by completing the square, giving the solutions in surd form.

13. $x^2 + 8x = 1$ **14.** $x^2 - 2x - 2 = 0$ **15.** $x^2 + x - 1 = 0$

16. $2x^2 + 2x = 1$ **17.** $x^2 + 3x + 1 = 0$ **18.** $2x^2 - x - 2 = 0$

19. $x^2 + 4x = 2$ **20.** $3x^2 + x - 1 = 0$ **21.** $2x^2 + 4x = 7$

22. $x^2 - x = 3$ **23.** $4x^2 + x - 1 = 0$ **24.** $2x^2 - 3x - 4 = 0$

The Formula for Solving a Quadratic Equation

Solving a quadratic equation by completing the square is rather tedious. If the method is applied to a general quadratic equation, a formula can be derived which can then be used to solve any particular equation.

Using a, b and c to represent any numbers we have the general quadratic equation

$$ax^2 + bx + c = 0$$

Using the method of completing the square for this equation gives

$$ax^2 + bx = -c$$

i.e. $$a\left(x^2 + \frac{b}{a}x\right) = -c \quad \Rightarrow \quad x^2 + \frac{b}{a}x = -\frac{c}{a}$$

$\therefore$ $$x^2 + \frac{b}{a}x + \left(\frac{b}{2a}\right)^2 = \left(\frac{b}{2a}\right)^2 - \frac{c}{a}$$

$\therefore$ $$\left(x + \frac{b}{2a}\right)^2 = \frac{b^2}{4a^2} - \frac{c}{a} = \frac{b^2 - 4ac}{4a^2}$$

$\Rightarrow$ $$x + \frac{b}{2a} = \pm\sqrt{\frac{(b^2 - 4ac)}{4a^2}} \quad \Rightarrow \quad x = -\frac{b}{2a} \pm \frac{\sqrt{(b^2 - 4ac)}}{2a}$$

i.e. $$x = \frac{-b \pm \sqrt{(b^2 - 4ac)}}{2a}$$

Example 3e

Find, by using the formula, the roots of the equation $2x^2 - 7x - 1 = 0$ giving them correct to 3 decimal places.

$$2x^2 - 7x - 1 = 0$$

Comparing with $ax^2 + bx + c = 0$ gives $a = 2$, $b = -7$, $c = -1$

$$x = \frac{-b \pm \sqrt{(b^2 - 4ac)}}{2a} = \frac{7 \pm \sqrt{\{49 - 4(2)(-1)\}}}{4}$$

Therefore, in surd form, $$x = \frac{7 \pm \sqrt{57}}{4}$$

Correct to 3 dp the roots are 3.637 and -0.137

EXERCISE 3e

Solve the equations by using the formula. Give the solutions in surd form.

1. $x^2 + 4x + 2 = 0$ 2. $2x^2 + x - 2 = 0$

3. $x^2 + 5x + 1 = 0$ 4. $2x^2 - x - 4 = 0$

5. $x^2 + 1 = 4x$ **6.** $2x^2 - x = 5$

7. $1 + x - 3x^2 = 0$ **8.** $3x^2 = 1 - x$

9. $5 + 2x = x^2$ **10.** $5x^2 - 1 = x$

Find, correct to 3 dp, the roots of the equations

11. $5x^2 + 9x + 2 = 0$ **12.** $2x^2 - 7x + 4 = 0$

13. $4x^2 - 7x - 1 = 0$ **14.** $3x = 5 - 4x^2$

15. $4x^2 + 3x = 5$ **16.** $1 = 5x - 5x^2$

17. $8x - x^2 = 1$ **18.** $x^2 - 3x = 1$

19. $2x^2 - x = 2$ **20.** $x^2 + 1 = 8x$

SIMULTANEOUS EQUATIONS

When only one unknown quantity has to be found, only one equation is needed to provide a solution.

If two unknown quantities are involved in a problem we need two equations connecting them. Then, between the two equations we can eliminate one of the unknowns, producing just one equation containing just one unknown. This is then ready for solution.

Linear Simultaneous Equations

There are two methods for reducing a pair of linear equations in two unknowns to one equation in one unknown.

The first method involves combining the left-hand sides and the right-hand sides of the two equations in the *same* way to produce another equality with just one unknown. This method is called *elimination* and is illustrated in first of the following worked examples.

Examples 3f

1. Solve the equations $x + 2y = 5$ and $3x + 2y = 11$

Subtracting the l.h.s. of the first equation from the l.h.s. of the second equation will eliminate the terms in y. To preserve the equality, the same process must be applied to the r.h.s. of the equations,

i.e. $(3x + 2y) - (x + 2y) = (11) - (5) \Rightarrow 2x = 6$

In practice it is easier to keep track of the process, and explain what is being done, when the equations are listed one below the other and numbered.

$$x + 2y = 5 \qquad\qquad [1]$$

$$3x + 2y = 11 \qquad\qquad [2]$$

$$[2] - [1] \quad\Rightarrow\quad 2x = 6 \;\Rightarrow\; x = 3$$

Now that the value of x is found, it can be substituted into either [1] or [2] to find the value of y.

Substituting 3 for x in [1] $\quad\Rightarrow\quad 3 + 2y = 5 \;\Rightarrow\; y = 1$

The solution should now be checked by substituting 3 for x and 1 for y in the *other* equation.

Check: When $x = 3$ and $y = 1$,   l.h.s. of [2] $\quad\Rightarrow\quad 9 + 2 = 11 = $ r.h.s.

In the second method we arrange one equation so that we can express one letter in terms of the other. This expression is then substituted for that letter in the other equation. This method is called *substitution* and is illustrated in the next worked example.

2. Solve the equations $y = 3x - 2$ and $x + 2y = 2$

$$y = 3x - 2 \qquad\qquad [1]$$

$$x + 2y = 2 \qquad\qquad [2]$$

[1] gives y in terms of x, so we will use it in this form.

Substituting $3x - 2$ for y in [2] $\quad\Rightarrow\quad x + 2(3x - 2) = 2$

$\Rightarrow \qquad\qquad\qquad\qquad\qquad\qquad x + 6x - 4 = 2$

$\Rightarrow \qquad\qquad\qquad\qquad\qquad\qquad 7x = 6 \;\Rightarrow\; x = \frac{6}{7}$

Substituting for x in [1] $\quad\Rightarrow\quad y = 3\left(\frac{6}{7}\right) - 2 \;\Rightarrow\; y = \frac{4}{7}$

Check: Using $x = \frac{6}{7}$ and $y = \frac{4}{7}$ in l.h.s. of [2] $\quad\Rightarrow\quad \frac{6}{7} + \frac{8}{7} = 2 = $ r.h.s.

Note that these equations can be solved by elimination. If this method is chosen it is sensible to rearrange the equations so that the letter terms and the number term are in the same order in each equation. In this example we could leave [1] as it is and rearrange [2] in the form $2y = -x + 2$.

Which method to use is a matter of choice. As a general guide, elimination gives a fairly direct solution when the coefficient of one letter is the same in both equations. That letter can then be eliminated by adding or subtracting the equations.

EXERCISE 3f

Solve the equations simultaneously for x and y, giving your answers in exact form.

1. $x + 3y = 0$
    $x - y = 4$

2. $3x - 5y = 7$
    $4x + 5y = -14$

3. $3x + 2y = 12$
    $3x - y = 9$

4. $2x + y = 7$
    $3x + 2y = 11$

5. $x + y = a$
    $x - y = b$

6. $y = 4 - x$
    $x = y - 6$

7. $\dfrac{x}{2} - y = 1$
    $x + y = 3$

8. $y = 6 - x$
    $2x + y = 8$

9. $x + \sqrt{2}y = 1$
    $x + \sqrt{3}y = 2$

10. $x - y = 5$
     $ax + by = 2$

11. $x = y + a$
     $y = mx + c$

12. $tx - t = 2y + t$
     $y = tx - c$

One Linear and One Quadratic Equation

If one of the equations is linear and the other is quadratic, one unknown can be isolated in the linear equation and then substituted in the quadratic equation.

Example 3g

Solve the equations $x - y = 2$
                           $2x^2 - 3y^2 = 15$

$$x - y = 2 \qquad [1]$$
$$2x^2 - 3y^2 = 15 \qquad [2]$$

Equation [1] is linear so we use it for the substitution, i.e. $x = y + 2$

Substituting $y + 2$ for x in [2] gives

$$2(y + 2)^2 - 3y^2 = 15$$
$$\Rightarrow \qquad 2(y^2 + 4y + 4) - 3y^2 = 15$$
$$\Rightarrow \qquad 2y^2 + 8y + 8 - 3y^2 = 15$$

Collecting terms on the side where y^2 is positive gives

$$0 = y^2 - 8y + 7$$
$$\Rightarrow \qquad 0 = (y - 7)(y - 1)$$
$$\therefore \qquad y = 7 \quad \text{or} \quad 1$$

Now we use $x = y + 2$ to find corresponding values of x

y	7	1
x	9	3

$\therefore$ either $\quad x = 9 \quad$ and $\quad y = 7$

or $\quad x = 3 \quad$ and $\quad y = 1$

Note that the values of x and y must be given in *corresponding pairs*.

It is incorrect to write the answer as $y = 7$ or 1 and $x = 9$ or 3

because $\begin{cases} y = 7 & \text{with} \quad x = 3 \\ y = 1 & \text{with} \quad x = 9 \end{cases}$ are *not* solutions

EXERCISE 3g

Solve the following pairs of equations.

1. $x^2 + y^2 = 5$
   $y - x = 1$

2. $y^2 - x^2 = 8$
   $x + y = 2$

3. $3x^2 - y^2 = 3$
   $2x - y = 1$

4. $y = 4x^2$
   $y + 2x = 2$

5. $y^2 + xy = 3$
   $2x + y = 1$

6. $x^2 - xy = 14$
   $y = 3 - x$

7. $xy = 2$
   $x + y - 3 = 0$

8. $2x - y = 2$
   $x^2 - y = 5$

9. $y - x = 4$
   $y^2 - 5x^2 = 20$

10. $x + y^2 = 10$
    $x - 2y = 2$

11. $4x + y = 1$
    $4x^2 + y = 0$

12. $3xy - x = 0$
    $x + 3y = 2$

13. $x^2 + 4y^2 = 2$
    $2y + x + 2 = 0$

14. $x + 3y = 0$
    $2x + 3xy = 1$

15. $3x - 4y = 1$
    $6xy = 1$

16. $x^2 + 4y^2 = 2$
    $x + 2y = 2$

17. $xy = 9$
    $x - 2y = 3$

18. $4x + y = 2$
    $4x + y^2 = 8$

19. $1 + 3xy = 0$
    $x + 6y = 1$

20. $x^2 - xy = 0$
    $x + y = 1$

21. $xy + y^2 = 2$
    $2x + y = 3$

Using Calculators and Computers

The methods for solving equations given in this chapter are algebraic. The advantage of these methods is that they can be done using pencil and paper and, more importantly, they give exact solutions when irrational numbers are involved. Further, when some of the number terms in an equation are unknown and represented by letters, algebraic methods produce solutions in terms of those letters, i.e. they give symbolic solutions.

Hand-held calculators are available that give numerical solutions, in decimals, for a variety of equations. The obvious limitation of these calculators is that they do not give solutions in surd form or in symbolic form; they are also not very cheap and not necessary for this level of work. However, if you are lucky enough to have one it is well worth learning how to use it because it does provide an easy way of *checking* results and can be used when exact answers are not needed. There are also computer programs that do give exact and symbolic solutions to equations, however these are not yet a realistic option for most people; they need powerful (i.e. expensive) computers and, apart from the expense, such computers are not yet small enough to carry around as easily as a calculator! If you have access to such a program, use it to solve some of the equations given earlier in this chapter.

There are methods other than algebraic, for finding numerical solutions to equations. These methods are mainly graphical and we will investigate them in a later chapter. (A graphics calculator is useful here, but again is not absolutely necessary.)

In the remaining section of this chapter, we use the symbolic solution of a quadratic equation to deduce some interesting and useful facts about the nature of the roots of such equations.

PROPERTIES OF THE ROOTS OF A QUADRATIC EQUATION

A number of interesting facts can be observed by examining the formula used for solving a quadratic equation, especially when it is written in the form

$$x = -\frac{b}{2a} \pm \frac{\sqrt{(b^2 - 4ac)}}{2a}$$

The Sum of the Roots

The separate roots are

$$-\frac{b}{2a} + \frac{\sqrt{(b^2 - 4ac)}}{2a} \quad \text{and} \quad -\frac{b}{2a} - \frac{\sqrt{(b^2 - 4ac)}}{2a}$$

When the roots are added, the terms containing the square root disappear giving

$$\text{sum of roots} = -\frac{b}{a}$$

This fact is very useful as a check on the accuracy of roots that have been calculated.

The Nature of the Roots

In the formula there are two terms. The first of these, $-\dfrac{b}{2a}$, can always be found for any values of a and b.

The second term however, i.e. $\dfrac{\sqrt{(b^2 - 4ac)}}{2a}$, is not so straightforward as there are three different cases to consider.

1) If $b^2 - 4ac$ is positive, its square root can be found and, whether it is a whole number, a fraction or a decimal, it is a number of the type we are familiar with – it is called a *real* number.

The two square roots, i.e. $\pm\sqrt{(b^2 - 4ac)}$ have different (or distinct) values, giving two different real values of x
So the equation has *two different real roots.*

2) If $b^2 - 4ac$ is zero then its square root also is zero and in this case
$x = -\dfrac{b}{2a} - \dfrac{\sqrt{(b^2 - 4ac)}}{2a}$ gives

$$x = -\frac{b}{2a} + 0 \quad \text{and} \quad x = -\frac{b}{2a} - 0$$

i.e. there is just one value of x that satisfies the equation.

An example of this case is $x^2 - 2x + 1 = 0$

From the formula we get $x = -\dfrac{(-2)}{2} \pm 0$

i.e. $x = 1$ or 1

By factorising we can see that there are two equal roots.

i.e. $(x - 1)(x - 1) = 0$

$\Rightarrow$ $x = 1$ or $x = 1$

This type of equation can be said to have a *repeated root*.

3) If $b^2 - 4ac$ is negative we cannot find its square root because there is no real number whose square is negative. In this case the equation has *no real roots*.

From these three considerations we see that the roots of a quadratic equation can be
be   either   real and different
         or   real and equal
         or   not real

and that it is the value of $b^2 - 4ac$ which determines the nature of the roots.

The expression '$b^2 - 4ac$' is called the *discriminant*.

Condition	Nature of Roots
$b^2 - 4ac > 0$	Real and different
$b^2 - 4ac = 0$	Real and equal
$b^2 - 4ac < 0$	Not real

Sometimes it matters only that the roots are real, in which case the first two conditions can be combined to give

if $b^2 - 4ac \geqslant 0$,   the roots are real.

Examples 3h

1. Determine the nature of the roots of the equation $x^2 - 6x + 1 = 0$

$x^2 - 6x + 1 = 0$

$a = 1, \quad b = -6, \quad c = 1$

$b^2 - 4ac = (-6)^2 - 4(1)(1) = 32$

$b^2 - 4ac > 0$ so the roots are real and different.

2. If the roots of the equation $2x^2 - px + 8 = 0$ are equal, find the value of p.

$$2x^2 - px + 8 = 0$$

$a = 2, \quad b = -p, \quad c = 8$

The roots are equal so $b^2 - 4ac = 0$,

i.e. $\quad (-p)^2 - 4(2)(8) = 0$

$\Rightarrow \qquad\qquad p^2 - 64 = 0$

$\Rightarrow \qquad\qquad p^2 = 64$

$\therefore \qquad\qquad p = \pm 8$

3. Prove that the equation $(k-2)x^2 + 2x - k = 0$ has real roots whatever the value of k

$$(k-2)x^2 + 2x - k = 0$$

$a = k - 2, \quad b = 2, \quad c = -k$

$b^2 - 4ac = 4 - 4(k-2)(-k)$

$\qquad\qquad = 4 + 4k^2 - 8k$

$\qquad\qquad = 4k^2 - 8k + 4$

$\qquad\qquad = 4(k^2 - 2k + 1) = 4(k-1)^2$

Now $(k-1)^2$ cannot be negative whatever the value of k, so $b^2 - 4ac$ cannot be negative. Therefore the roots are always real.

EXERCISE 3h

Without solving the equation, write down the sum of its roots.

1. $x^2 - 4x - 7 = 0$ 2. $3x^2 + 5x + 1 = 0$

3. $2 + x - x^2 = 0$ 4. $3x^2 - 4x - 2 = 0$

5. $x^2 + 3x + 1 = 0$ 6. $7 + 2x - 5x^2 = 0$

Without solving the equation, determine the nature of its roots.

7. $x^2 - 6x + 4 = 0$ 8. $3x^2 + 4x + 2 = 0$

9. $2x^2 - 5x + 3 = 0$ 10. $x^2 - 6x + 9 = 0$

11. $4x^2 - 12x - 9 = 0$ 12. $4x^2 + 12x + 9 = 0$

13. $x^2 + 4x - 8 = 0$ 14. $x^2 + ax + a^2 = 0$

15. $x^2 - ax - a^2 = 0$ 16. $x^2 + 2ax + a^2 = 0$

17. If the roots of $3x^2 + kx + 12 = 0$ are equal, find k

18. If $x^2 - 3x + a = 0$ has equal roots, find a

19. The roots of $x^2 + px + (p - 1) = 0$ are equal. Find p

20. Prove that the roots of the equation $kx^2 + (2k + 4)x + 8 = 0$ are real for all values of k

21. Show that the equation $ax^2 + (a + b)x + b = 0$ has real roots for all values of a and b

22. Find the relationship between p and q if the roots of the equation $px^2 + qx + 1 = 0$ are equal

MIXED EXERCISE 3

In Questions 1 to 6, solve the equation giving x in an exact form.

1. $3x - 2 = 5(2 - x)$ 2. $\dfrac{5}{x} = 4$ 3. $\dfrac{1}{x + 1} = \dfrac{3}{x}$

4. $2x - \sqrt{3} = x + \sqrt{2}$ 5. $\dfrac{1}{x - \sqrt{2}} = \sqrt{2} - 1$ 6. $ax + b = 2x - c$

In questions 7 to 12, solve the equations for x and y.

7. $x - y = 2$ 8. $5x - 2y = 4$ 9. $x - 3y = 7$
    $2x + y = 4$ $3x - 2y = 1$ $2x + y = 9$

10. $x = y(\sqrt{2} - 1)$ 11. $y = ax + b$ 12. $x - ay = 2$
       $x - y = 2$ $2y = 3 - x$ $x + by = c$

In each Question from 13 to 22

(a) write down the value of $-b/a$
(b) use any suitable method to find the roots of the equation, giving any irrational roots in surd form.
(c) find the sum of the roots and check that it is equal to the answer to (a).

13. $x^2 - 5x - 6 = 0$ 14. $x^2 - 6x - 5 = 0$

15. $2x^2 + 3x = 1$ 16. $5 - 3x^2 = 4x$

17. $x(2 - x) = 1$ 18. $4x^2 - 3 = 11x$

19. $(x - 1)(x + 2) = 1$ 20. $x^2 + 4x + 4 = 16$

21. $x^2 + 2x = 2$ **22.** $2(x^2 + 2) = x(x - 4)$

In Questions 23 to 28, solve the equations giving *all possible* solutions.

23. $x(x - 2) = 0$ **24.** $x(x + 3) = 4$

25. $x^2 + x + 5 = x(2x + 5)$ **26.** $x^2 + 5x + 2 = 2(2x + 1)$

27. $x(x - 5) = 2(x + 5)$ **28.** $2x(x + 3) = x(x - 1) + 7$

In Questions 29 and 30 solve the pair of equations. (Choose your substitution carefully, to keep the amount of squaring to a minimum.)

29. $2x^2 - y^2 = 7$ **30.** $2x = y - 1$
$x + y = 9$ $x^2 - 3y + 11 = 0$

31. Use the formula to solve the equation $3x^2 - 17x + 10 = 0$

  (a) Are the roots of the equation rational or irrational?

  (b) What does your answer to (a) tell you about the left-hand side of the equation?

32. Determine the nature of the roots of the equations

  (a) $x^2 + 3x + 7 = 0$ (b) $3x^2 - x - 5 = 0$

  (d) $ax^2 + 2ax + a = 0$ (d) $2 + 9x - x^2 = 0$

33. For what values of p does the equation $px^2 + 4x + (p - 3) = 0$ have equal roots?

34. Show that the equation $2x^2 + 2(p + 1)x + p = 0$ always has real roots.

35. The equation $x^2 + kx + k = 1$ has equal roots. Find k.

36. The solution to the equation given below contains some errors. Find them and explain why they are errors. Give the correct solution.

1:      $\dfrac{x}{x - 1} = \dfrac{x^2}{2}$

2: $\Rightarrow$ $\dfrac{1}{x - 1} = \dfrac{x}{2}$

3: $\Rightarrow$ $\dfrac{2}{2x - 2} = x$

4: $\Rightarrow$ $2 = 2x^2 - 2$

5: $\Rightarrow$ $2x^2 = 4 \Rightarrow x = \pm\sqrt{2}$

CONSOLIDATION A

SUMMARY

Terms and Coefficients

In an algebraic expression, terms are separated by plus or minus signs. An individual term is identified by the combination of letters involved. The coefficient of a term is the number in the term, e.g. $\textcircled{2}x^2y$

Expansion of Brackets

Important results are
$$(ax + b)^2 = a^2x^2 + 2abx + b^2$$
$$(ax - b)^2 = a^2b^2 - 2abx + b^2$$
$$(ax + b)(ax - b) = a^2x^2 - b^2$$

Pascal's Triangle

$$
\begin{array}{ccccccccc}
 & & & & 1 & & 1 & & \\
 & & & 1 & & 2 & & 1 & \\
 & & 1 & & 3 & & 3 & & 1 \\
 & 1 & & 4 & & 6 & & 4 & & 1 \\
\end{array}
$$

The 1st, 2nd, 3rd, ... rows in this array give the coefficients in the expansion of $(1 + x)^1$, $(1 + x)^2$, $(1 + x)^3$, ...

Indices

$$a^n \times a^m = a^{n+m}$$
$$a^n \div a^m = a^{n-m}$$
$$(a^n)^m = a^{nm}$$
$$\sqrt[n]{a} = a^{1/n}$$
$$a^0 = 1$$
$$a^{n/m} = \sqrt[m]{a^n} = (\sqrt[m]{a})^n$$

Linear Equations

A linear equation involves linear expressions only.

Simultaneous Equations

When two unknown quantities have to be found, two equations connecting them are required.

These two equations are solved simultaneously by reducing them to one equation involving one unknown quantity. This can be achieved by elimination or by substitution.

Quadratic Equations

The general quadratic equation is $ax^2 + bx + c = 0$

The roots of this equation can be found by

        factorising when this is possible,

or   completing the square,

or   by using the formula $x = \dfrac{-b \pm \sqrt{(b^2 - 4ac)}}{2a}$

When $b^2 - 4ac > 0$, the roots are real and different.

When $b^2 - 4ac = 0$, the roots are real and equal.

When $b^2 - 4ac < 0$, the roots are not real.

MULTIPLE CHOICE EXERCISE A

TYPE I

1. The roots of the equation $x^2 - 3x + 2 = 0$ are

   A 2, 1   **B** $-2, -1$ **C** $-3, 2$ **D** $0, \frac{2}{3}$ **E** not real

2. The coefficient of xy in the expansion of $(x - 3y)(2x + y)$ is

   A 1   **B** 6   **C** 5   **D** 0   **E** -5

3. $\dfrac{1 - \sqrt{2}}{1 + \sqrt{2}}$ is equal to

   A 1   **B** -1 **C** $3 - \sqrt{2}$ **D** $1 - \frac{2}{3}\sqrt{2}$ **E** $2\sqrt{2} - 3$

4. Expanding $(1 + \sqrt{2})^3$ gives

   A $3 + 3\sqrt{2}$ **C** $1 + 3\sqrt{2}$ **E** $1 + 2\sqrt{2}$
   B $7 + 5\sqrt{2}$ **D** $3 + \sqrt{6}$

5. If $x^2 + px + 6 = 0$ has equal roots and $p > 0$ then p is

   A $\sqrt{48}$ **B** 0 **C** $\sqrt{6}$ **D** 3 **E** $\sqrt{24}$

6. If $x^2 + 4x + p \equiv (x + q)^2 + 1$, the values of p and q are

   A $p = 5, q = 2$ **D** $p = -1, q = 5$
   B $p = 1, q = 2$ **E** $p = 0, q = -1$
   C $p = 2, q = 5$

7. $\dfrac{p^{-1/2} \times p^{3/4}}{p^{-1/4}}$ simplifies to

   A 1 **B** $p^{-1/2}$ **C** $p^{3/4}$ **D** p **E** $p^{1/2}$

8. In the expansion of $(a - 2b)^3$ the coefficient of b^2 is

   A $-2a^2$ **B** $-8a$ **C** $12a$ **D** $-4a$ **E** -12

TYPE II

9. If $y = 2x - 1$ and $xy = 3$ then there is only one value of y that satisfies both equations.

10. When $(3 - 5x)^4$ is expanded the coefficient of x^4 is 1.

11. $x^2 - 2x + 2 = (x - 1)^2 + 1$

12. $2x^2 + 3x - 2$ has a factor $(x + 2)$

13. $\dfrac{2\sqrt{3} - 2}{2\sqrt{3} + 2}$ is an irrational number

14. If $x - a$ is a factor of $x^2 + px + q$, the equation $x^2 + px + q = 0$ has a root equal to a.

15. $10x^6 \div x^{1/2} = 10x^3$

16. In the expansion of $(1 + x)^6$ the coefficient of x is 6

17. Values of A and B can be found such that
$$3x^2 - 2x + 7 \equiv (x - A)^2 + B$$

MISCELLANEOUS EXERCISE A

1. Find the value of x when (a) $3^x = 3\sqrt{3}$ (b) $5^x = 125\sqrt{5}$

2. Express $(2x+1)(x-2) - 3$ as a product of linear factors. (UCLES)$_s$

3. Solve the simultaneous equations
$$x + y = 2, \qquad x^2 + 2y^2 = 11 \qquad\qquad \text{(UCLES)}_s$$

4. Find the value of c for which $x = 2$ is a root of the equation
$3x^2 - 4x + c = 0$

5. Find both solutions of the equation $x + 1 = \dfrac{20}{x}$ (OCSEB)

6. Express $3x^2 + 6x - 4$ in the form $a[(x+b)^2 + c]$

7. Find the values of p and q for which
$$x^2 - 4x + p \equiv (x - q)^2 + 4$$

8. Find all the solutions that satisfy both the equations $x^2 + y^2 = 9$ and
$x - 2y + 3 = 0$

9. Find the value of k for which the equation $x^2 - 9x + k = 0$ has equal roots.

10. Find the value of a for which $x - 1$ is a factor of $2x^2 - 3x + a$

11. Show that the roots of the equation $x^2 + kx - 2 = 0$ are real and different
for all values of k

12. Determine the range of values of k for which the quadratic equation
$x^2 + 2kx + 4k - 3 = 0$ has two real and distinct roots. (WJEC)

13. Factorise (a) $x^3 + 7x$ (b) $x^2 - y^2$ (c) $(a - 2)^2 - b^2$

14. Simplify $\dfrac{2 + \sqrt{5}}{2 - \sqrt{5}} - \sqrt{5}$

15. Solve the equation $5(x - 3) - 2(x - 4) = 9$

16. Solve the equations $2x - y + a = 3$ and $x + 2y - 2a = 4$ giving answers in
terms of a.

17. Find the values of p and q for which
$$2x^2 + px + 3 = 2[(x - 1)^2 + q]$$
Hence show that there are no real values of x for which
$$2[(x - 1)^2 + q] = 0$$

18. Find the value of a for which $x = 1$ is a solution of the equation
$2x^2 - 3x + a = 0$

19. Simplify (a) $(\sqrt{3} - 2)(4 - 3\sqrt{3}) + \sqrt{3}$ (b) $(x - 2)(2x + 1) + x(3x + 4)$

20. Find, correct to 2 decimal places, the solutions of the equations
$$2x + 3y = 5$$
$$x^2 + 2xy + y^2 = 6$$

21. Find the values of a, b and c for which
$$(x - a)(x^2 - 4x + 5) \equiv x^3 + bx^2 + cx - 15$$

22. Solve the equation $2(5x - x^2 - 1) + 2(x^3 - 7x + 1) = 0$

23. Simplify (a) $\dfrac{x^2 - 6x + 5}{2x^2 - 2}$ (b) $\dfrac{2\sqrt{7}}{\sqrt{7} + 3} + \dfrac{\sqrt{7}}{3 - \sqrt{7}}$

24. Find the values of a and b for which
$$(2x - a)^3 = 8x^3 + bx^2 + 6a^2x - 27$$

25. Given that k is a real constant such that $0 < k < 1$, show that the roots of
the equation
$$kx^2 + 2x + (1 - k) = 0$$
are   (a) always real
        (b) always negative.   (ULEAC)

26. A cuboid measures 27 cm by 20 cm by 10 cm. Find the maximum absolute
error when these measurements are used to calculate the volume of the cuboid if
the measurements are given

(a) correct to the nearest centimetre

(b) rounded down to the nearest centimetre.

27. Given that $a = \sqrt{2}$, $b = \sqrt{3}$ and $c = \sqrt{5}$, express the following in terms
of a, b and c only.

(a) $\sqrt{24} - \sqrt{75}$ (b) $\dfrac{\sqrt{12} - \sqrt{36}}{\sqrt{18} + \sqrt{27}}$

28. Solve the equations for x, giving solutions in terms of π.

(a) $\pi x^2 = 25 - 5x$ (b) $\dfrac{9\pi}{x} + \dfrac{1}{\pi^2(x - 1)} = 0$

29. (a) Write down the values of $\sqrt{2}$ and $\sqrt{3}$ correct to 3 significant figures. Use
these values to find $2\sqrt{3} + 3\sqrt{2}$ as a decimal.

(b) Use your calculator to find $2\sqrt{3} + 3\sqrt{2}$ correct to 3 significant figures.

(c) What is the percentage error in the answer for part (a)?

30. Find S and T in terms of m_1 and m_2 when

$$2S - m_1 T = m_2 \quad \text{and} \quad T - \frac{S}{m_2} = 3m_1$$

31. Find the value of n for which

    (a) $x^n = \sqrt{x^3}$ (b) $x^n = x^3 \div x^{-4}$ (c) $x^n = (x^{1/3})^2$ (d) $x^n = 1$

32. By expressing $\frac{1}{32}$ as a power of 2, solve the equation $2^{3x+1} = \frac{1}{32}$

CHAPTER 4

STRAIGHT LINE GEOMETRY

PROOF

This chapter contains some geometric facts and definitions that will be needed later in this course.

Up to now, many rules have been based on investigating a few particular cases. For example the reader may have accepted that the sum of the interior angles in any triangle is 180°, only because the measured angles in some specific triangles had this property. This fact may be reinforced by our not being able to find a triangle whose angles have a different sum but that does not rule out the possibility that such a triangle exists.

It is no longer satisfactory to assume that a fact is *always* true without *proving that it is*, because as a mathematics course progresses, one fact is often used to produce another. Hence it is very important to distinguish between a 'fact' that is assumed from a few particular cases and one that has been *proved* to be true, as results deduced from an assumption cannot be reliable.

A proof deals with a general case, e.g. a triangle in which the sides and angles are not specified. The formal statement of a proved result is called a *theorem*.

PROOF THAT THE ANGLES OF A TRIANGLE ADD UP TO 180°

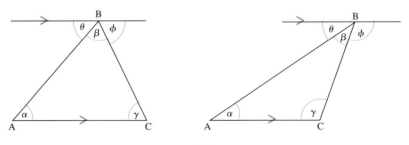

Let ABC be any triangle.

Draw a straight line through B parallel to AC.

Using the notation on the diagram, and the fact that alternate angles are equal, we have

$$\alpha = \theta \ \text{ and } \ \gamma = \phi$$

At B, $\theta + \beta + \phi = 180°$ (angles on a straight line)

∴ $\alpha + \beta + \gamma = 180°$

We have proved the general case and hence can now be certain that, for *all* triangles *the angles of a triangle add up to 180°*

Proof that in any triangle, an exterior angle is equal to the sum of the two interior opposite angles

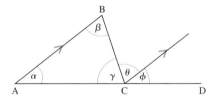

Let ABC be any triangle, with AC produced to D.

Draw a line through C parallel to AB.

Using the notation on the diagram,

$$\theta = \beta \ \text{ (alternate } \angle s) \ \text{ and } \ \phi = \alpha \ \text{ (corresponding } \angle s)$$

∴ $\theta + \phi = \beta + \alpha$

i.e. $\angle BCD = \angle BAC + \angle CBA$

DEFINITIONS

Division of a Line in a Given Ratio

A point P is said to divide a line AB *internally* if P is between A and B.

Further, if P divides AB internally in the ratio $p : q$, then

$\qquad$ AP : PB $= p : q$

If a point P is on AB produced, or on BA produced, then P is said to divide a line AB *externally*.

Further, if P divides AB externally in the ratio $p : q$ then AP : PB $= p : q$

Examples 4a

1. A line AB of length 15 cm is divided internally by P in the ratio 2 : 3. Find the length of PB.

As P divides AB internally in the ratio 2 : 3, P is between A and B and is nearer to A than to B.

AB is divided into 5 portions of which PB is 3 portions. If one portion is x cm, then

$\qquad$ AB $= 5x = 15$

$\Rightarrow$ $\qquad\qquad x = 3$

$\therefore$ $\qquad$ PB $= 3x = 9$

i.e. $\qquad$ PB is 9 cm long.

2. A line AB of length 15 cm is divided externally by P in the ratio 2 : 3. Find the length of PB.

As AP : PB = 2 : 3, P is nearer to A than to B, so P is on BA produced.

$$AB = 3x - 2x = 15$$

$$\Rightarrow \qquad\qquad x = 15$$

$$\therefore \qquad PB = 3x = 45$$

i.e.   PB is 45 cm long.

EXERCISE 4a

1. A line AB of length 12 cm is divided internally by P in the ratio 1 : 5. Find the length of AP.

2. A line AB of length 18 cm is divided externally by P in the ratio 5 : 6. Find the length of AP.

3. The point D divides a line PQ internally in the ratio 3 : 4. If PQ is of length 35 cm, find the length of DQ.

4. A line LM is divided externally by a point T in the ratio 5 : 7. Find the length of MT if (a) LM = 24 cm (b) LM = $2x$ cm.

5. P is a point on a line AB of length 12 cm and AP = 5 cm. Find the ratio in which P divides AB.

6. AB is a line of length 16 cm and M is a point on AB produced such that AM = 24 cm. Find the ratio in which M divides AB.

7. AB is a line of length x units and P is a point on AB such that AP is of length y units. Find, in terms of x and y, the ratio in which P divides AB.

8. PQ is a line of length a units. It is divided externally in the ratio $n : m$ by a point L, where $n > m$. Find, in terms of a, n and m, the length of QL.

9. ST is a line of length a units and L is a point on ST produced such that TL is of length b units. Find, in terms of a and b, the ratio in which L divides ST.

THE INTERCEPT THEOREM

A straight line drawn parallel to one side of a triangle divides the other two sides in the same ratio.

i.e.

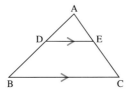

if DE is parallel to BC then AD/DB = AE/EC

Proof

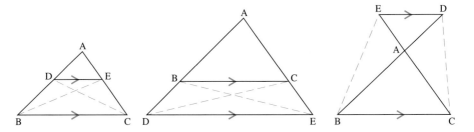

Triangle ABC is any triangle and DE is parallel to BC.

Three different positions for DE must be considered, as shown in the diagrams. The proof that follows applies to all three cases. This proof uses the fact that the area of a triangle can be found by multiplying half the base by the height.

$\qquad$ Area $\triangle$DEC $=$ area $\triangle$DEB

$\qquad\qquad\qquad$ ($\triangle$s have same base DE and equal heights)

$\therefore\qquad$ area $\triangle$AED : area $\triangle$DEB $=$ area $\triangle$AED : area $\triangle$DEC

Now$\quad$ area $\triangle$AED : area $\triangle$DEB $=$ AD : DB

$\qquad\qquad\qquad$ ($\triangle$s have the same heights)

and$\quad$ area $\triangle$AED : area $\triangle$DEC $=$ AE : EC

$\qquad\qquad\qquad$ ($\triangle$s have the same heights)

Therefore AD : DB $=$ AE : EC

The converse of this theorem is also true, i.e. if a line divides two sides of a triangle in the same ratio, then it is parallel to the third side of the triangle.

PYTHAGORAS' THEOREM AND ITS CONVERSE

Pythagoras' theorem is familiar and very useful. Here is a reminder.

In any right-angled triangle, the square on the hypotenuse is equal to the sum of the squares on the other two sides.

i.e.

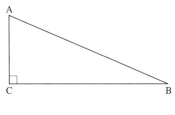

$$AB^2 = AC^2 + BC^2$$

The converse of this theorem is less well known, but equally useful. It states that

**if the square on one side of a triangle is equal to the sum of the squares on the other two sides,
then the angle opposite the first side is a right-angle.**

EXERCISE 4b

1. In triangle ABC, DE is parallel to BC.
   If AD = 2 cm, DB = 3 cm
   and AE = 2.5 cm, find EC.

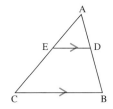

2. In triangle LMN, ST is parallel to MN.
   If LM = 9 cm, LS = 4 cm
   and LT = 2 cm, find
   (a) TN (b) LN.

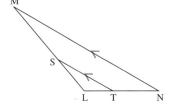

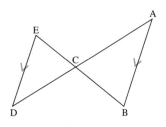

3. In the diagram, ED is parallel to AB.
    AC = 3 cm, CE = 1.5 cm
    and CD = 2 cm. Find BC.

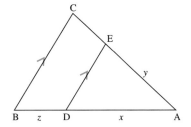

4. Using the measurements given in
    the diagram find the length of
    CE in terms of x, y and z

5. ABC is a triangle and a line PQ is drawn parallel to BC, but on the opposite
    side of A from BC. PQ cuts BA produced at P and cuts CA produced at Q.
    AC = 2.5 cm, AQ = 1 cm and AP = 0.5 cm. Find
    (a) the length of AB (b) the ratio in which P divides AB.

6. Determine whether or not a triangle is right-angled if the lengths of its sides are

    (a) 3, 5, 4   (b) 2, 1, $\sqrt{3}$ (c) 2, 2, $\sqrt{8}$
    (d) $5x$, $12x$, $13x$ (e) 7, 5, 12   (f) $\sqrt{2}$, $\sqrt{3}$, 1

SIMILAR TRIANGLES

If one triangle is an enlargement of another triangle, then the two triangles are
similar.

This means that the three angles of one triangle are equal to the three angles of
the other triangle *and* that the corresponding sides of the two triangles are in the
same ratio.

However, to prove that triangles are similar it is necessary only to show that *one*
of these conditions is satisfied because the other one follows, i.e.

**if two triangles contain the same angles then their corresponding
sides are in the same ratio.**

Proof

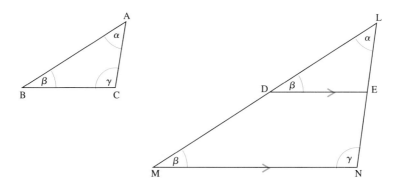

The angles of triangle ABC are equal to the angles of triangle LMN.

It follows that we can find a point D on LM such that LD = AB and a point E on LN such that LE = AC.

Joining DE, it is clear that △LDE and △ABC are identical, i.e. congruent.

∴   ∠ABC = ∠LDE

These are corresponding angles with respect to DE and MN.

∴   DE is parallel to MN

The intercept theorem then tells us that DE divides LM and LN in the same ratio, i.e.

        LD : LM = LE : LN

⇒   AB : LM = AC : LN

The converse of this theorem is also true, i.e.

if two triangles are such that their corresponding sides are in the same ratio, then corresponding pairs of angles are equal.

It is left to the reader to prove this in Question 14 in Exercise 4d.

SIMILAR FIGURES

Two figures are similar if one figure is an enlargement of the other. This means that their corresponding sides are in the same ratio and that the angles in one figure are equal to the corresponding angles in the other figure.

To show that figures other than triangles are similar, both the side and the angle property have to be proved. In the case of triangles, we have seen that it is necessary only to show that one of these conditions is satisfied to prove the triangles similar, because the other condition follows, i.e.

two triangles are similar if we can show
either that the angles of the triangles are equal
or that the corresponding sides of the triangles are in the same ratio.

CONGRUENT FIGURES

Two figures are congruent if they are identical in shape and size. This means that corresponding angles and corresponding sides are equal. In the case of triangles, it is not necessary to prove that all three angles and all three sides are equal to show that triangles are congruent.

Two triangles are congruent if, in both triangles

two sides and the included angle
are equal

or

two angles and a corresponding
side are equal

or

all three sides are equal

or

two sides are equal and one
angle is a right-angle

EXERCISE 4c

These diagrams are not drawn to scale. State whether the two triangles are congruent, not congruent but similar, neither congruent nor similar.

1.
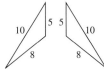

2.
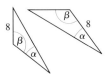

3.
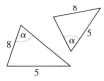

4.

5.

6.

7.

8.

9.

THE ANGLE BISECTOR THEOREM

Another useful fact concerning triangles and ratios is

> **the line bisecting an angle of a triangle divides the side opposite to that angle in the ratio of the sides containing the angle.**

e.g. if AD bisects $\angle A$, then $BD : DC = AB : AC$

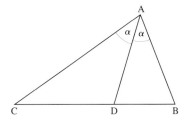

Proof

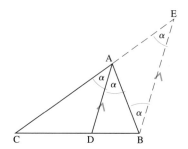

In △ABC, AD bisects the angle at A.

Drawing BE parallel to DA to cut CA produced at E, we have

        ∠BEA = ∠DAC (corresponding angles)

and   ∠EBA = ∠BAD (alternate angles)

∴   △BEA is isosceles ⇒ EA = AB

In △BCE the intercept theorem gives

        BD : DC = EA : AC

⇒   BD : DC = AB : AC

ALTITUDES AND MEDIANS

We end this chapter with a couple of definitions.

The line drawn from a vertex of a triangle, perpendicular to the opposite side, is called an *altitude*, for example

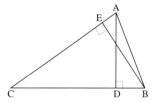

      AD is the altitude through A,

and BE is the altitude through B.

A *median* of a triangle is the line joining a vertex to the midpoint of the opposite side, for example

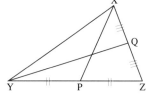

      XP is the median through X,

and YQ is the median through Y.

Example 4d

In △ABC, AB = 4 cm, AC = 3 cm and ∠A = 90°. The bisector of ∠A cuts BC at D. Find the length of BD.

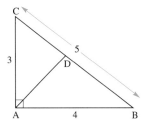

From Pythagoras' theorem, BC = 5 cm.

From the angle bisector theorem, BD : DC = AB : AC = 4 : 3

$$\therefore \qquad\qquad\qquad\qquad BD : BC = 4 : 7$$

$$\Rightarrow \qquad\qquad\qquad\qquad BD = \tfrac{4}{7} \times 5$$

$$= 2\tfrac{6}{7}$$

$\therefore \qquad$ BD is $2\tfrac{6}{7}$ cm long.

EXERCISE 4d

1. XYZ is a triangle with a right angle at X, XW is the altitude from X. Show that triangles XYZ, XWZ and XYW are all similar.

2. In △ABC, AB = 6 cm, AC = 5 cm and BC = 7 cm. BD is the median from B to AC and BE is the bisector of ∠B to AC. Find the length of DE.

3. Triangles PQR and XYZ are such that ∠P = ∠X and ∠Q = ∠Z. XY = 3 cm, YZ = 4 cm, PQ = 7 cm and PR = 12 cm. Find the lengths of XZ and QR.

4.

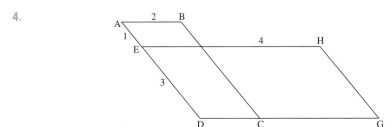

ABCD and EDGH are parallelograms. Prove that they are similar.

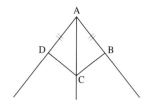

5. The diagram shows a design for a roof section. AC bisects angle DAB and AD = AB. By showing that triangles ADC and ABC are congruent, show that DC = CB.

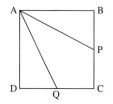

6. ABCD is a square. P is the midpoint of BC and Q is the midpoint of DC. By finding a pair of congruent triangles, show that AP = AQ.

7. Draw a rectangle ABCD and mark E, the midpoint of AB. Join DE and CE. Prove that DE and CE are the same length.

8. PQR is a triangle in which PQ = 4 cm, PR = 3 cm and QR = 6 cm. T is a point outside the triangle on the side of QR, and ∠RQT = ∠PRQ and ∠QRT = ∠QPR. Find the lengths of QR and RT.

9. ABC is any triangle and equilateral triangles ABD and ACE are drawn on the sides AB and AC respectively. The bisector of ∠BAC meets BC at F such that BF : FC = 3 : 2. Find the ratio of the areas of the two equilateral triangles.

10. D and E are two points on the side BC of △ABC such that AD is the bisector of ∠BAC and AE is an altitude of the triangle. If ∠ACB = 40° and ∠ABC = 60° find ∠DAE.

11. In triangle ABC, AB = 24 cm, BC = 7 cm and AC = 25 cm. Show that △ABC is right-angled. The bisector of ∠BAC meets BC at D. Find the lengths of BD, DC and AD.

12. AB and CD are two lines that intersect at E. AC and DB are parallel. Show that triangles ACE and EDB are similar.

13. Triangle ABC has a right angle at B and BE is an altitude of the triangle. BC = 5 cm and BE = 4 cm. Calculate the length of EC and of AC.

14. (Proof that if the corresponding sides of two triangles are in the same ratio, then the triangles contain the same angles.)

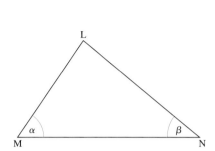
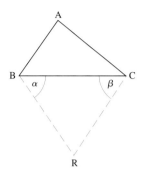

Triangles LMN and ABC are such that LM : AB = MN : BC = LN : AC.
R is a point such that ∠CBR = ∠LMN and ∠BCR = ∠LNM.
Prove that LM : BR = LM : AB and hence that BR = AB.
Similarly show that CR = AC. *Hence* show that △LMN and △ABC have equal angles.

15. A quadrilateral is such that both pairs of opposite sides are parallel.
Without assuming any other facts about the shape of this quadrilateral, prove that

(a)   the opposite sides are equal in length

(b)   the diagonals bisect each other.

16. A triangle has two equal sides. Prove that the angles opposite these sides are equal.

17. A quadrilateral ABCD is such that AB = AD and CB = CD. Prove that the diagonal AC bisects the angles at A and C.

CHAPTER 5

COORDINATE GEOMETRY

LOCATION OF A POINT IN A PLANE

Graphical methods lend themselves particularly well to the investigation of the geometrical properties of many curves and surfaces. At this stage we will restrict ourselves to rectilinear plane figures (i.e. two-dimensional figures bounded by straight lines). To begin, we need a simple and unambiguous way of describing the position of a point on a graph.

Consider the problem of describing the location of a city, London say.

There are many ways in which this can be done, but they all require reference to at least one known place and known directions. This is called a *system* or *frame of reference*. Within this frame of reference, two measurements are needed to locate the city precisely. These measurements are called coordinates.

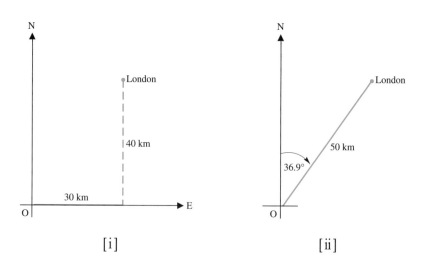

[i]   [ii]

The position of London is described in two alternative ways in the diagrams above.

In [i] the frame of reference comprises a fixed point O and the directions due east and due north from O. The coordinates of London are 30 km east of O and 40 km north of O.

In [ii] the frame of reference comprises a fixed point O and the direction due north from O. The coordinates of London are 50 km from O and a bearing of 036.9°

The system used at this level for graphical work is based on the first of the two practical systems described above.

CARTESIAN COORDINATES

This system of reference uses a fixed point O, called *the origin*, and a pair of perpendicular lines through O. One of these lines is drawn horizontally and is called the *x*-axis. The other line is drawn vertically and is called the *y*-axis.

The coordinates of a point P are the directed distances of P from O parallel to the axes.

A positive coordinate is a distance measured in the positive direction of the axis and a negative coordinate is a distance in the opposite direction.

The coordinates are given as an ordered pair (a, b) with the x-*coordinate* or *abscissa* first and the y-*coordinate* or *ordinate* second.

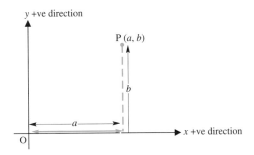

The diagram overleaf represents the points whose Cartesian coordinates are (7, 4) and $(-3, -1)$

These points are referred to in future simply as the points (7, 4) and $(-3, -1)$

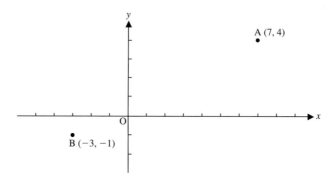

EXERCISE 5a

1. Represent on a diagram the points whose coordinates are
   (a) (1, 6)   (b) (0, 5)   (c) (−4, 0)   (d) (−3, −2)   (e) (3, −4)

2. Two adjacent corners of a square are the points (3, 5) and (3, −1). What
   could the coordinates of the other two corners be?

3. The two opposite corners of a square are (−2, −3) and (3, 2). Write down
   the coordinates of the other two corners.

COORDINATE GEOMETRY

Coordinate geometry is the name given to the graphical analysis of geometric
properties. For this analysis we need to refer to three types of points:

1) fixed points whose coordinates are known, e.g. the point (1, 2)

2) fixed points whose coordinates are not known numerically. These points are
referred to as (x_1, y_1), (x_2, y_2), ... etc. or (a, b), etc.

3) points which are not fixed. We call these general points and we refer to them
as (x, y), (X, Y), etc.

It is conventional to use the letters A, B, C, ... for fixed points and the letters P,
Q, R, ... for general points.

It is also conventional to graduate the axes using identical scales. This avoids
distorting the shape of figures.

THE LENGTH OF A LINE JOINING TWO POINTS

Consider the line joining the points $A(1, 2)$ and $B(3, 4)$

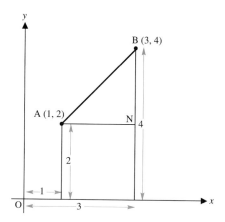

The length of the line joining A and B can be found by using Pythagoras' theorem, i.e.

$$AB^2 = AN^2 + BN^2$$
$$= (3-1)^2 + (4-2)^2$$
$$= 8$$

Therefore $AB = \sqrt{8} = 2\sqrt{2}$

In the same way the length of the line joining any two points $A(x_1, y_1)$ and $B(x_2, y_2)$ can be found.

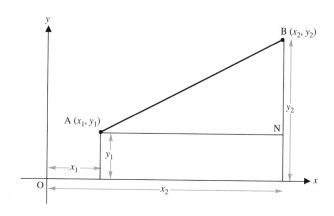

From the diagram, $AB^2 = AN^2 + BN^2$

$$= (x_2 - x_1)^2 + (y_2 - y_1)^2$$

$\Rightarrow \qquad AB = \sqrt{[(x_2 - x_1)^2 + (y_2 - y_1)^2]}$

i.e.   **the length of the line joining $A(x_1, y_1)$ to $B(x_2, y_2)$ is given by**
$$AB = \sqrt{[(x_2 - x_1)^2 + (y_2 - y_1)^2]}$$

This formula still holds when some, or all, of the coordinates are negative. This is illustrated in the next worked example.

Examples 5b

1. Find the length of the line joining $A(-2, 2)$ to $B(3, -1)$

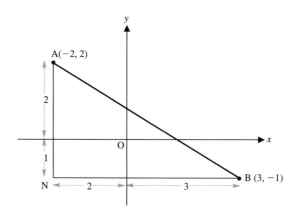

$AB = \sqrt{[(x_2 - x_1)^2 + (y_2 - y_1)^2]}$

$\quad = \sqrt{[(3 - \{-2\})^2 + (-1 - 2)^2]}$

$\quad = \sqrt{(5^2 + \{-3\}^2)}$

$\quad = \sqrt{34}$

From the diagram, $BN = 3 + 2 = 5$

and $\qquad AN = 2 + 1 = 3$

$\Rightarrow \qquad AB^2 = 5^2 + 3^2 = 34$

$\Rightarrow \qquad AB = \sqrt{34}$

This confirms that the formula used above is valid when some of the coordinates are negative.

EXERCISE 5b

1. Find the length of the line joining
   (a) $A(1, 2)$ and $B(4, 6)$ (b) $C(3, 1)$ and $D(2, 0)$
   (c) $J(4, 2)$ and $K(2, 5)$

2. Find the length of the line joining
   (a) $A(-1, -4)$, $B(2, 6)$ (b) $S(0, 0)$, $T(-1, -2)$
   (c) $E(-1, -4)$, $F(-3, -2)$

3. Find the distance from the origin to the point $(7, 4)$

4. Find the length of the line joining the point $(-3, 2)$ to the origin.

5. Find the length of the line drawn from the point $(4, -8)$ to the origin.

6. Show, by using Pythagoras' Theorem, that the lines joining $A(1, 6)$, $B(-1, 4)$ and $C(2, 1)$ form a right-angled triangle.

7. A, B and C are the points $(7, 3)$, $(-4, 1)$ and $(-3, -2)$ respectively.
   (a) Show that $\triangle ABC$ is isosceles.
   (b) Find the area of $\triangle ABC$ by enclosing it in a rectangle, or otherwise.

8. The vertices of a triangle are $A(0, 2)$, $B(1, 5)$ and $C(-1, 4)$
   Find the perimeter of the triangle.

9. Show that the lines OA and OB are perpendicular where A and B are the points $(4, 3)$ and $(3, -4)$ respectively.

10. The point $A(0, 4)$ is the centre of a circle whose radius is 5. Show that the point $B(4, 1)$ is on the circumference of the circle.

11. Plot the points $A(5, 4)$ and $B(7, 8)$. Write down the coordinates of the midpoint D of the line joining A and B.

12. Repeat question 11 with the points $A(-1, 2)$ and $B(3, -4)$.

13. By looking at the results of questions 11 and 12, state, in each case, the relationship between the x-coordinate of D and the x-coordinates of A and B. Does the same relationship exist between the y-coordinate of D and the y-coordinates of A and B?

14. P is the point (a, b) and Q is the point (c, d). Write down, in terms of a, b, c and d, the coordinates of the midpoint of the line joining P and Q.

The Midpoint of a Line joining Two Points

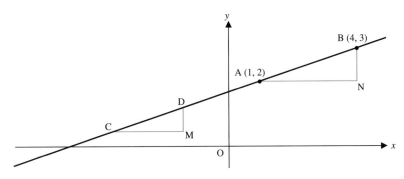

Questions 11 to 14 in Exercise 5b show that the coordinates of M are the average of the coordinates of A and B.

GRADIENT

The gradient of a straight line is a measure of its slope with respect to the *x*-axis. Gradient is defined as

the increase in *y* divided by the increase in *x* between one point and another point on the line.

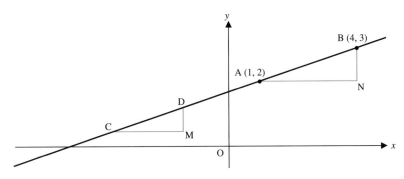

Consider the line passing through A(1, 2) and B(4, 3).

From A to B,   the increase in *y* is 1
                 the increase in *x* is 3

Therefore the gradient of AB is $\frac{1}{3}$.

Now NB measures the increase in the *y*-coordinate and AN measures the increase in the *x*-coordinate, so the gradient can be written as $\dfrac{NB}{AN}$.

If C and D are any other two points on the line then $\triangle ABN$ and $\triangle CDN$ are similar,

so $\dfrac{NB}{AN} = \dfrac{MD}{CM} = \dfrac{1}{3}$

i.e.   the gradient of a line may be found from *any* two points on the line.

Now consider the line through the points $A(2, 3)$ and $B(6, 1)$

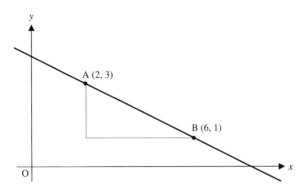

Moving from A to B

$$\frac{\text{increase in } y}{\text{increase in } x} = \frac{-2}{4} = -\frac{1}{2}$$

Alternatively, moving from B to A

$$\frac{\text{increase in } y}{\text{increase in } x} = \frac{2}{-4} = -\frac{1}{2}$$

This shows that it does not matter in which order the two points are considered, provided that they are considered in the *same* order when calculating the increases in x and in y.

From these two examples we see that the gradient of a line may be positive or negative.

A positive gradient indicates an 'uphill' slope with respect to the positive direction of the x-axis, i.e. the line makes an acute angle with the positive sense of the x-axis.

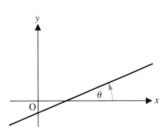

A negative gradient indicates a 'downhill' slope with respect to the positive direction of the x-axis, i.e. the line makes an obtuse angle with the positive sense of the x-axis.

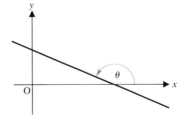

In general,

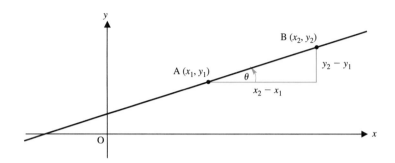

the gradient of the line passing through $A(x_1, y_1)$ and $B(x_2, y_2)$ is

$$\frac{\text{the increase in } y}{\text{the increase in } x} = \frac{y_2 - y_1}{x_2 - x_1}$$

As the gradient of a straight line is the increase in y divided by the increase in x from one point on the line to another,

gradient measures the increase in y per unit increase in x, i.e. the rate of increase of y with respect to x.

PARALLEL LINES

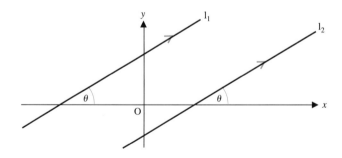

If l_1 and l_2 are parallel lines, they are equally inclined to the positive direction of the x-axis, i.e.

parallel lines have equal gradients.

PERPENDICULAR LINES

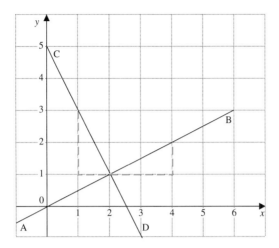

The diagram shows a line AB whose gradient is 2, and a line CD which is perpendicular to AB.

From the diagram we can see that the gradient of CD is $-\frac{1}{2}$.

This illustrates the general fact that

> **if a line AB has gradient m,**
> **any line perpendicular to AB has gradient** $-\dfrac{1}{m}$

from which we see that

> **the product of the gradients of perpendicular lines is** -1

Example 5c

Determine, by comparing gradients, whether the following three points are collinear (i.e. lie on the same straight line); $A(\frac{2}{3}, 1)$, $B(1, \frac{1}{2})$, $C(4, -4)$

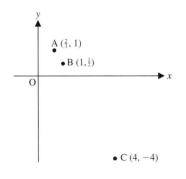

The gradient of AB is
$$\frac{1 - \frac{1}{2}}{\frac{2}{3} - 1} = -\frac{3}{2}$$

The gradient of BC is
$$\frac{-4 - \frac{1}{2}}{4 - 1} = -\frac{3}{2}$$

As the gradients of AB and BC are the same, A, B and C are collinear.

The diagram, although not strictly necessary, gives a check that the answer is reasonable.

EXERCISE 5c

1. Find the gradient of the line through the pair of points.

   (a) $(0, 0), (1, 3)$

   (b) $(1, 4), (3, 7)$

   (c) $(5, 4), (2, 3)$

   (d) $(-1, 4), (3, 7)$

   (e) $(-1, -3), (-2, 1)$

   (f) $(-1, -6), (0, 0)$

   (g) $(-2, 5), (1, -2)$

   (h) $(3, -2), (-1, 4)$

   (i) $(-4, -3), (-1, -8)$

   (j) $(h, k), (0, 0)$

2. Determine whether the given points are collinear.

   (a) $(0, -1), (1, 1), (2, 3)$

   (b) $(0, 2), (2, 5), (3, 7)$

   (c) $(-1, 4), (2, 1), (-2, 5)$

   (d) $(0, -3), (1, -4), (-\frac{1}{2}, -\frac{5}{2})$

   (e) $(-1, -2), (0, 4), (2, 8)$

3. Determine whether AB and CD are parallel, perpendicular or neither.

   (a) $A(0, -1), B(1, 1), C(1, 5), D(-1, 1)$

   (b) $A(1, 1), B(3, 2), C(-1, 1), D(0, -1)$

   (c) $A(3, 3), B(-3, 1), C(-1, -1), D(1, -7)$

   (d) $A(2, -5), B(0, 1), C(-2, 2), D(3, -7)$

   (e) $A(2, 6), B(-1, -9), C(2, 11), D(0, 1)$

   (f) $A(-1, -2), B(0, 4), C(5, 2), D(4, 8)$

PROBLEMS IN COORDINATE GEOMETRY

This chapter ends with a miscellaneous selection of problems on coordinate geometry. A clear and reasonably accurate diagram showing all the given information will often suggest the most direct method for solving a particular problem.

Example 5d

The vertices of a triangle are the points $A(2, 4)$, $B(1, -2)$ and $C(-2, 3)$ respectively. The point $H(a, b)$ lies on the altitude through A. Find a relationship between a and b

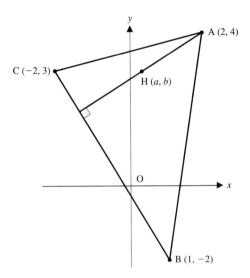

As H is on the altitude through A, AH is perpendicular to BC.

The gradient of AH is $\dfrac{4-b}{2-a}$,

the gradient of BC is $\dfrac{3-(-2)}{-2-1} = -\dfrac{5}{3}$

The product of the gradients of perpendicular lines is -1

Therefore $\qquad \left[\dfrac{4-b}{2-a}\right]\left[-\dfrac{5}{3}\right] = -1$

$\Rightarrow \qquad\qquad \dfrac{-20+5b}{6-3a} = -1$

$\Rightarrow \qquad\qquad 5b = 3a + 14$

EXERCISE 5d

1. $A(1, 3)$, $B(5, 7)$, $C(4, 8)$, $D(a, b)$ form a rectangle ABCD. Find a and b.

2. The triangle ABC has its vertices at the points $A(1, 5)$, $B(4, -1)$ and $C(-2, -4)$

   (a) Show that $\triangle ABC$ is right-angled.

   (b) Find the area of $\triangle ABC$.

3. Show that the point $(-\frac{32}{3}, 0)$ is on the altitude through A of the triangle whose vertices are A$(1, 5)$, B$(1, -2)$ and C$(-2, 5)$

4. Show that the triangle whose vertices are $(1, 1)$, $(3, 2)$, $(2, -1)$ is isosceles.

5. Find, in terms of a and b, the length of the line joining (a, b) and $(2a, 3b)$

6. The point $(1, 1)$ is the centre of a circle whose radius is 2. Show that the point $(1, 3)$ is on the circumference of this circle.

7. A circle, radius 2 and centre the origin, cuts the x-axis at A and B and cuts the positive y-axis at C. *Prove* that $\angle$ACB $= 90°$

8. Show that $\triangle$OCD is isosceles if O is the origin, C is the point (p, q) and D is the point (q, p)

9. The point (a, b) is on the circumference of the circle of radius 3 whose centre is at the point $(2, 1)$. Find a relationship between a and b

10. The vertices of a triangle are at the points A$(a, 0)$, B$(0, b)$ and C(c, d) and $\angle$B $= 90°$. Find a relationship between a, b, c and d

11. A point P(a, b) is equidistant from the y-axis and from the point $(4, 0)$. Find a relationship between a and b

12. Show that, for all values of a, the point P$(5 - a, a)$ is equidistant from the points A$(1, 2)$ and B$(3, 4)$.

13. A line with gradient 1 passes through the point $(2, 1)$. Give the coordinates of the points where this line cuts the axes. Hence find the area enclosed by this line and the axes.

CHAPTER 6

BASIC TRIGONOMETRY

TRIGONOMETRY AND RIGHT-ANGLED TRIANGLES

Basic trigonometry is concerned with the relationships between the angles and the sides in a right-angled triangle. It has wide-ranging practical applications, some of which we look at later in this chapter.

Consider this triangle which has a right angle at B. The side opposite the right angle is always called the hypotenuse. The two shorter sides can be identified with respect to ∠A as shown in the diagram.

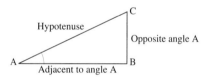

All triangles with a right angle and an angle equal to ∠A are similar.

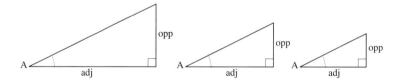

For all these triangles the ratio $\dfrac{\text{side opp A}}{\text{side adj A}}$ is the same.

The Tangent of An Angle

For any given angle, x, in a right-angled triangle, the ratio $\dfrac{\text{side opp } x}{\text{side adj } x}$ has been calculated. It is called the *tangent of angle x* and is usually abbreviated to tan x.

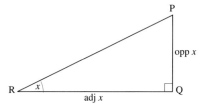

$$\tan x = \frac{\text{side opp } x}{\text{side adj } x} = \frac{PQ}{RQ}$$

The value of the tan of any angle can be found from a scientific calculator. First make sure that the calculator is in 'degree mode' (calculators can give angles in other units).

Calculators vary; to find $\tan 20°$, either press 20 | tan | (this works on most calculators) or press | tan | 20 | EXE |

Conversely, if the value of the tan ratio is known, the angle can be found, e.g. if $\tan x = 0.5$, enter 0.5 | tan⁻¹ | (or | tan⁻¹ | 20 | EXE |) to find x.

The worked examples show how tangents of angles can be used to find the size of an angle or the length of a side in a right-angled triangle. It is a good habit to label the sides of the triangle 'opp' and 'adj' to the angle being used.

Examples 6a

1. In $\triangle ABC$, $\angle B = 90°$, $BC = 3.2$ cm, $AB = 5.3$ cm. Find the size of $\angle A$.

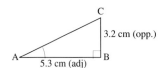

$$\tan A = \frac{\text{opp}}{\text{adj}} = \frac{3.2}{5.3} = 0.6307\ldots$$

$\therefore \angle A = 31.1°$ correct to 1 decimal place.

2. In $\triangle PQR$, $\angle P = 90°$, $\angle Q = 32°$, and $PQ = 7.2$ m. Find the length of PR.

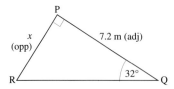

$$\tan 32° = \frac{\text{opp}}{\text{adj}} = \frac{x}{7.2}$$

$\therefore \quad x = 7.2 \times \tan 32° = 4.499\ldots$

$PQ = 4.50$ m correct to 3 s.f.

3. Find the length of the side marked x.

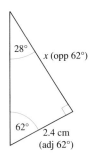

The unknown side is adjacent to the angle 28°. Using this angle gives the equation

$$\tan 28° = \frac{2.4}{x}$$

An equation with the unknown on the top is easier to solve, and we can get such an equation in this case by using the third angle in the triangle.

$$\tan 62° = \frac{x}{2.4} \quad \Rightarrow \quad x = 2.4 \times \tan 62° = 4.513\ldots$$

The length of the side marked x is 4.51 cm (3 sf)

When an answer has been found, it is important to check that it is reasonable in relation to the problem. In the case of triangles, all that is needed is a quick check that the basic geometry is right, i.e. that the shortest side is opposite the smallest angle and, in a right-angled triangle, that the hypotenuse is the longest side.

Throughout this chapter, give angles corrected to 1 decimal place and lengths corrected to 3 significant figures, unless otherwise instructed.

EXERCISE 6a

Write down, correct to 4 significant figures, the tan of the angle.

1. 28°   **2.** 83.5°   **3.** 2.4°   **4.** 57.8°

Write down the angle whose tan is

5. 0.5972   **6.** $\frac{2}{3}$ **7.** $\frac{5}{4}$ **8.** $\frac{\sqrt{3}}{2}$

9. Find the value of θ when
  (a) $\tan\theta = 1.5$ (b) $\tan\theta = 5/2$ (c) $\tan\theta = \sqrt{5/4}$

The following diagrams are not drawn to scale. The lengths of sides are in metres and the angles are in degrees.
In questions 10 to 15, find the size of the angle marked θ.

10.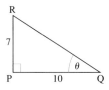

11.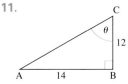

12.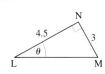

13.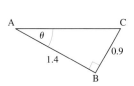

14.

15.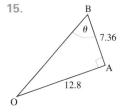

In questions 16 to 21, find the length of the side marked x.

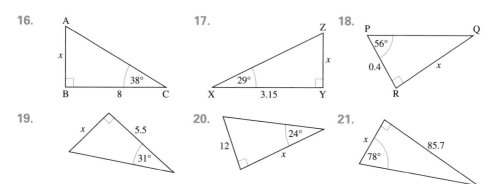

16.   17.   18.

19.   20.   21.

Sine and Cosine

Two more ratios can be found using different pairs of sides of a right-angled triangle.

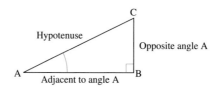

In a right-angled triangle,

the ratio $\dfrac{\text{side opposite } \angle A}{\text{hypotenuse}}$ is called the sine of angle A and is written as sin A

the ratio $\dfrac{\text{side adjacent to } \angle A}{\text{hypotenuse}}$ is called the cosine of angle A and this is written as cos A

In any right-angled triangle

$$\sin A = \frac{\text{opp A}}{\text{hyp}} \qquad \cos A = \frac{\text{adj A}}{\text{hyp}} \qquad \tan A = \frac{\text{opp A}}{\text{adj A}}$$

The calculations are done in a way similar to those involving the tan of an angle.

Example 6b

Triangle $\triangle$LMN has a right angle at L, LM $= 6.4$ m and $\angle$N $= 37°$. Find the length of MN.

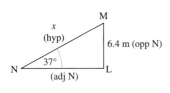

With respect to N, we know the opposite side and we want the hypotenuse. The trig ratio using these two sides is sine.

$$\sin 37° = \frac{\text{opp}}{\text{hyp}} = \frac{6.4}{x}$$

$$\Rightarrow \quad x \sin 37° = 6.4$$

$$\Rightarrow \quad x = \frac{6.4}{\sin 37°} = 10.63\ldots$$

$\therefore$ LN $= 10.6$ m correct to 3 sf.

When the hypotenuse needs to be found, there is no way of avoiding x being in the denominator.

EXERCISE 6b

Give trig ratios correct to 4 significant figures.

1. Find the sine of (a)   76°   (b)   22.5°

2. Find the cosine of (a)   84°   (b)   5°

3. Find (a)   sin 71°   (b)   sin 18°   (c)   sin 27°
         (d)   cos 12°   (e)   cos 82°   (f)   cos 30°

4. Find θ when (a) $\sin \theta = 0.467$ (b) $\sin \theta = 1/\sqrt{5}$
                  (c) $\cos \theta = 4/7$ (d) $\cos \theta = 2/3$

5. Find the size of angle C using the sine ratio and check your answer using the cosine ratio.

(a)   (b)   (c)

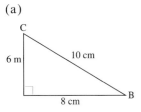

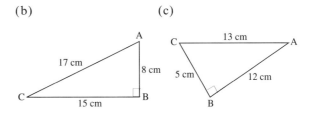

6. State whether the sine or cosine ratio should be used to find the size of angle θ, and find θ.

(a)   (b)   (c)

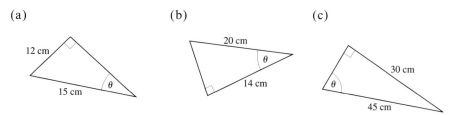

For questions 7 to 12 find the length marked x or the angle marked θ; you may need to use the sine, cosine or tangent ratio.

7.   8.   9.

10.   11.   12.

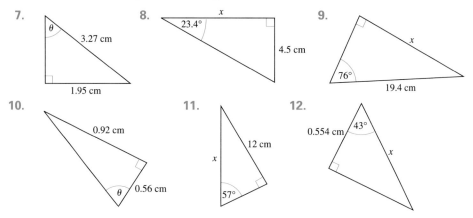

Applications

There are many applications, e.g. in surveying, where trigonometry can be used to find distances and/or angles. Remember that, at this stage, trig can be used only in right-angled triangles. If there is no right-angled triangle, the diagram needs to be broken down to give such a triangle.

Examples 6c

1. Find the length of the side marked x.

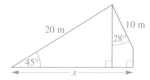

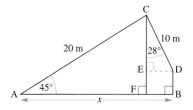

This diagram is easier to work with when it is lettered. To find AB, we see that AB = AF + FB. AF can be found from $\triangle$AFC and, by adding ED to the diagram and using $\triangle$CED, ED and hence FB can be found.

In $\triangle$AFC, $\qquad \dfrac{AF}{20} = \cos 45°$

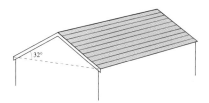

$\Rightarrow \qquad\qquad AF = 20 \times \cos 45° = 14.142\ldots$

In $\triangle$CED, $\qquad \dfrac{ED}{10} = \sin 28°$

$\Rightarrow \qquad\qquad ED = 10 \times \sin 28° = 4.694\ldots$

$AB = AF + FB = AF + ED = 18.836\ldots$

The length of the side marked x is 18.8 m (3 sf)

Note that intermediate results should not be corrected but kept live (in the memory of a calculator). Alternatively, the calculation can be done in one stage using $AB = AF + FB = 20 \cos 45 + 10 \sin 28$.

2. **The cross-section of this roof is an isosceles triangle. The tiled sides of the roof are inclined at 32° to the horizontal and the narrow end of the roof is 8 metres wide at the base. Find the height of the roof.**

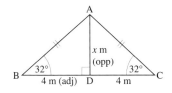

To use trigonometry in a problem such as this one, the first step is to produce a simple diagram showing triangles and to put the given information on this diagram. As we know that the cross-section is an isosceles triangle, we will show this in the diagram.

To use trigonometry in a problem such as this one, the first step is to produce a simple diagram showing triangles and to put the given information on this diagram. As we know that the cross-section is an isosceles triangle, we will show this in the diagram.

Any isosceles triangle can be divided by a line down the middle into two equal-sized right-angled triangles. Adding this information to the diagram, we see that AD represents the height of the roof.

Using $\triangle$ADB, $\qquad \tan 32° = \dfrac{x}{4} \quad \Rightarrow \quad x = 4 \tan 32° = 2.499\ldots$

The answer must now be translated into words to give a solution to the problem posed.

The height of the roof is 2.50 metres correct to 3 sf.

EXERCISE 6c

1. Find the length of the side marked x and/or the size of the angle marked θ in the following diagrams.

(a)

(b)

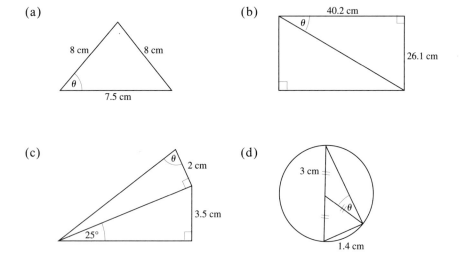

(c)

(d)

(e)

(f)

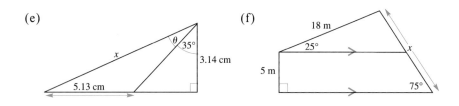

2. From the point F, which is level with the base, C, of a vertical cliff and 45 m away from it, the angle F is measured as 15°. Find the height of the cliff.

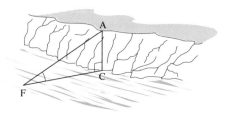

A big wheel at a funfair has a radius of 10 metres. When the chair is at A, it is 4 metres above the diameter PQ, find the angle between AO and PQ.

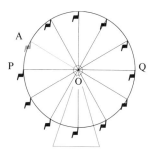

4. Find the angle made by the line through the points A(2, 4), B(−1, 2) with the x-axis.

5. To find the height of a vertical tower, a point A is taken on the ground such that A is level with the base B of the tower. The distance AB is measured and found to be 35 metres. The angle between the ground and the line joining A to the top of the tower C is also measured and found to be 56°. Use this information to find the height of the tower.

6. A plane flies 150 km north-east from Gatwick and then changes course and flies 100 km in the direction 30° east of due south. How far is it now from Gatwick?

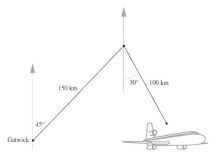

Exact Values For Trig Ratios

When a trig ratio is given as a fraction, the lengths of two of the sides of the right-angled triangle can be marked. Then the third side can be calculated by using Pythagoras' theorem, hence the other trig ratios can also be expressed as fractions.

Example 6d

Given that $\sin A = \frac{3}{5}$ find $\cos A$ and $\tan A$

Because $\sin A = \dfrac{\text{opp}}{\text{hyp}}$, we can draw a right-angled triangle with the side opposite to angle A of length 3 units and a hypotenuse of length 5 units.

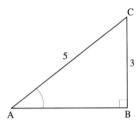

Applying Pythagoras' theorem to $\triangle ABC$ gives

$$(AB)^2 + (BC)^2 = (AC)^2$$

i.e. $\qquad (AB)^2 + 3^2 = 5^2$

$\Rightarrow \qquad\qquad (AB)^2 = 25 - 9 = 16$

$\Rightarrow \qquad\qquad\quad AB = 4$

Then $\qquad \cos A = \dfrac{\text{adj}}{\text{hyp}} = \dfrac{4}{5}$

and $\qquad \tan A = \dfrac{\text{opp}}{\text{adj}} = \dfrac{3}{4}$

THE TRIG RATIOS OF 30°, 45°, 60°

The sine, cosine and tangent of 30°, 45°, and 60°, can be expressed exactly in surd form and are worth remembering.

This triangle shows that $\qquad\qquad$ $\mathbf{\sin\ 45° = \dfrac{1}{\sqrt{2}}}$

$$\mathbf{\cos\ 45° = \dfrac{1}{\sqrt{2}}}$$

$$\mathbf{\tan\ 45° = 1}$$

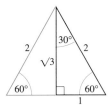

And this triangle gives
$$\sin 60° = \frac{\sqrt{3}}{2}, \quad \sin 30° = \frac{1}{2}$$
$$\cos 60° = \frac{1}{2}, \quad \cos 30° = \frac{\sqrt{3}}{2}$$
$$\tan 60° = \sqrt{3}, \quad \tan 30° = \frac{1}{\sqrt{3}}$$

EXERCISE 6d

If any of the square roots in this exercise are not integers, leave them in surd form.

1. If $\tan A = \frac{12}{5}$ find $\sin A$ and $\cos A$

2. Given that $\cos X = \frac{4}{5}$ find $\tan X$ and $\sin X$

3. If $\sin P = \frac{40}{41}$ find $\cos P$ and $\tan P$

4. $\operatorname{Tan} A = 1$ Find $\sin A$ and $\cos A$

5. If $\cos Y = \frac{2}{3}$ find $\sin Y$ and $\tan Y$

6. Given that $\sin A = \frac{1}{2}$ what is $\cos A$? Use your calculator to find the size of angle A

7. If $\sin X = \frac{7}{25}$ and $\tan X = \frac{7}{?}$ find $\cos X$

8. If $\sin X = \frac{3}{5}$ find $\cos X$ and hence calculate $\cos^2 X + \sin^2 X$

9. Repeat Question 8 with $\sin X = \frac{1}{2}$

10. If $\sin X = \frac{p}{q}$ find $\cos X$ in terms of p and q
    Hence find the value of $\sin^2 X + \cos^2 X$

11. Find, in surd form, the value of
    (a) $\sin 45° + \sin 60°$ (b) $\cos 30° + \cos 60°$ (c) $\dfrac{\tan 45°}{1 - \cos 45°}$

12. The length of the side of a square picture frame is measured as 4.5 m correct to 2 significant figures. The length of the diagonals is calculated using trigonometry. Find, correct to 3 significant figures, the maximum absolute error in the calculated length if cos 45° is taken as (a) $1/\sqrt{2}$ (b) 0.71

13. A flagpole AB is 12 m high. Two ties go from the top, A, of the pole to points C and D on level ground such that B, C and D are collinear. From C, the angle of elevation of A is 28° and from D, the angle of elevation of A is 47°. Find the distance between C and D.

14. A yacht sails from Cowes on a bearing of N 60° W for 500 m. It then changes course and sails 200 m on a bearing N 20° E and then anchors. How far from Cowes is the yacht now?

*15. Starting with a right-angled triangle with one angle of 45°, and using the angle bisector theorem, find exact forms for the values of sin 22.5°, cos 22.5° and tan 22.5°.

*16. A sphere whose radius is 5 cm fits inside a hollow cone so that the top of the sphere is 18 cm from the vertex of the cone. Find the radius of the circle where the cone touches the sphere.

*17. For a survey to find the depth of a gravel pit, the angle was measured between the horizontal and a line joining the measuring point to a marker C at the top of the pit. This was done from two points A and B distant 25 m apart as shown. Find the vertical depth of the pit.

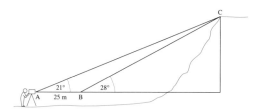

CHAPTER 7

STRAIGHT LINES

THE MEANING OF EQUATIONS

The Cartesian frame of reference provides a means of defining the position of any point in a plane. This plane is called the xy-plane.

In general x and y are independent variables. This means that they can each take any value independently of the value of the other unless some restriction is placed on them.

Consider the case when the value of x is restricted to 2

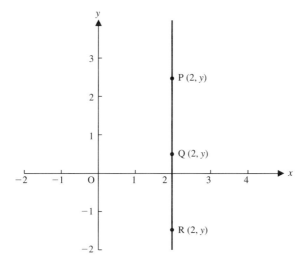

As the value of y is not restricted, the condition above gives a set of points which form a straight line parallel to the y-axis and passing through P, Q and R as shown. Therefore the condition that x is equal to 2, i.e. $x = 2$, defines the line through P, Q and R, i.e.

in the context of the xy-plane, the equation $x = 2$ defines the line shown in the diagram. Further, $x = 2$ is called *the equation of this line* and we can refer briefly to *the line $x = 2$*

Now consider the set of points for which the condition is $x > 2$

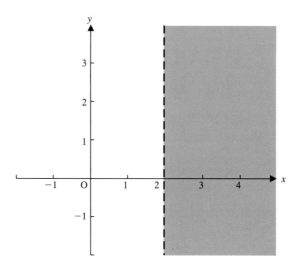

All the points to the right of the line $x = 2$ have an x-coordinate that is greater than 2

So the inequality $x > 2$ defines the shaded region of the xy-plane shown above.

Similarly, the inequality $x < 2$ defines the region left unshaded in the diagram.

Note that the region defined by $x > 2$ does not include the line $x = 2$

> **When a region does not include a boundary line this is drawn as a broken line. When the points on a boundary *are* included in a region, this boundary is drawn as a solid line.**

Example 7a

Draw a sketch of the region of the xy-plane defined by the inequalities
$0 \leqslant x \leqslant 2$ and $0 < y < 4$

The relationship $0 \leqslant x \leqslant 2$ contains two inequalities which must be considered separately, i.e. $x \geqslant 0$ and $x \leqslant 2$. Similarly, $0 < y < 4$ contains two relationships, i.e. $y > 0$ and $y < 4$. The required region is found by considering each inequality in turn and shading the unwanted region. This leaves the required region clear.

$x \geqslant 0$ defines both the line $x = 0$
(i.e. the y-axis) and the region to the
right of the y-axis $(x > 0)$ so we
shade the region to the left of the
y-axis.

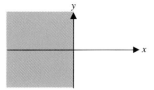

$x \leqslant 2$ defines both the line $x = 2$
and the region to the left of the line.

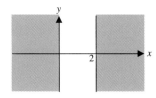

$y > 0$ defines the region above the
x-axis, but does not include the x-axis.

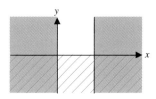

$y < 4$ defines the region below the
line $y = 4$ but does not include the
line $y = 4$

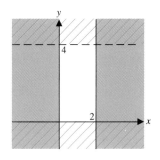

Therefore the unshaded region in the bottom figure, including the solid boundary
lines but not the broken ones, is the set of points that satisfy all the given
inequalities.

Note that in this book, we use the convention of shading unwanted regions when dealing with more than
one inequality in a plane. This convention is not universal and some problems ask for such a diagram to
be drawn with the required region shaded. In this case we recommend that the reader follows the
procedure used in the worked example and then redraws the diagram to shade the required region.

EXERCISE 7a

1. Draw a sketch showing the lines defined by the equations $x = 5$, $x = -3$, $y = 0$, $y = 6$

2. Draw a sketch showing the lines defined by the equations $y = -3$, $y = -10$, $x = 7$, $x = -5$

   Draw a sketch showing the region of the xy-plane defined by the following inequalities.

3. $x > 3$ 4. $y \leqslant 2$ 5. $x \geqslant -8$

6. $0 < x < 5$ 7. $-1 < y < 4$ 8. $-2 \leqslant x < 2$

9. $0 \leqslant x \leqslant 5$, $-1 \leqslant y \leqslant 3$ 10. $x \leqslant -3$, $x \geqslant 4$, $y < -2$

THE EQUATION OF A STRAIGHT LINE

A straight line may be defined in many ways; for example, a line passes through the origin and has a gradient of $\frac{1}{2}$.

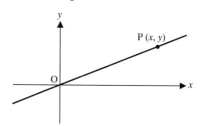

The point $P(x, y)$ is on this line if and only if the gradient of OP is $\frac{1}{2}$

In terms of x and y, the gradient of OP is $\frac{y}{x}$, so the statement above can be written in the form

$$P(x, y) \text{ is on the line if and only if } \frac{y}{x} = \frac{1}{2}, \text{ i.e. } 2y = x$$

Therefore the coordinates of points on the line satisfy the relationship $2y = x$, and the coordinates of points that are not on the line do not satisfy this relationship.

$2y = x$ is called the equation of the line.

The equation of a line (straight or curved) is a relationship between the x and y coordinates of all points on the line and which is not satisfied by any other point in the plane.

Examples 7b

1. Find the equation of the line through the points $(1, -2)$ and $(-2, 4)$.

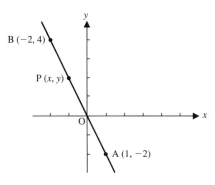

$P(x, y)$ is on the line if and only if the gradient of PA is equal to the gradient of AB (or PB).

The gradient of PA is $\dfrac{y - (-2)}{x - 1} = \dfrac{y + 2}{x - 1}$

The gradient of AB is $\dfrac{-2 - 4}{1 - (-2)} = -2$

Therefore the coordinates of P satisfy the equation $\dfrac{y + 2}{x - 1} = -2$

i.e. $\quad y + 2x = 0$

Consider the more general case of the line whose gradient is m and which cuts the y-axis at a directed distance c from the origin. Note that c is called the *intercept* on the y-axis.

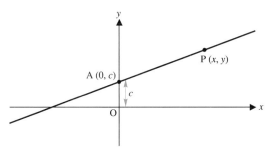

Now $P(x, y)$ is on this line if and only if the gradient of AP is m

Therefore the coordinates of P satisfy the equation $\dfrac{y - c}{x - 0} = m$

i.e. $\quad y = mx + c$

This is the *standard form* for the equation of a straight line.

An equation of the form $y = mx + c$ represents a straight line with gradient m and intercept c on the y-axis.

Because the value of m and/or c may be fractional, this equation can be rearranged and expressed as $ax + by + c = 0$, i.e.

$$ax + by + c = 0$$
where a, b and c are constants, is the equation of a straight line.

Note that in this form c is *not* the intercept.

Examples 7b (continued)

2. Write down the gradient of the line $3x - 4y + 2 = 0$ and find the equation of the line through the origin which is perpendicular to the given line.

Rearranging $3x - 4y + 2 = 0$ in the standard form gives

$$y = \tfrac{3}{4}x + \tfrac{1}{2}$$

Comparing with $y = mx + c$ we see that the gradient (m) of the line is $\tfrac{3}{4}$ (and the intercept on the y-axis is $\tfrac{1}{2}$).

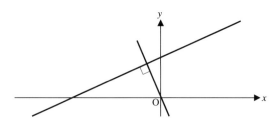

The gradient of the perpendicular line is given by $-\dfrac{1}{m} = -\dfrac{4}{3}$ and it passes through the origin so the intercept on the y-axis is 0

Therefore its equation is $y = -\tfrac{4}{3}x + 0$

$\Rightarrow$ $4y + 3x = 0$

3. Sketch the line $x - 2y + 3 = 0$

This line can be located accurately in the xy-plane when we know two points on the line. We will use the intercepts on the axes as these can be found easily (i.e. $x = 0 \Rightarrow y = \frac{3}{2}$ and $y = 0 \Rightarrow x = -3$).

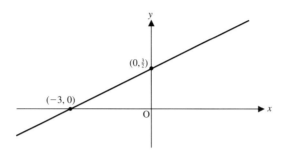

Notice that the diagrams in the worked examples are sketches, not accurate plots, but they show reasonably accurately the position of the lines in the plane.

EXERCISE 7b

1. Write down the equation of the line through the origin and with gradient

(a) 2   (b) -1 (c) $\frac{1}{3}$ (d) $-\frac{1}{4}$ (e) 0   (f) ∞

Draw a sketch showing all these lines on the same set of axes.

2. Write down the equation of the line passing through the given point and with the given gradient.

(a) $(0, 1)$, $\frac{1}{2}$ (b) $(0, 0)$, $-\frac{2}{3}$ (c) $(-1, -4)$, 4

Draw a sketch showing all these lines on the same set of axes.

3. Write down the equation of the line passing through the points

(a) $(0, 0)$, $(2, 1)$ (b) $(1, 4)$, $(3, 0)$ (c) $(-1, 3)$, $(-4, -3)$

4. Write down the equation of the line passing through the origin and perpendicular to

(a) $y = 2x + 3$ (b) $3x + 2y - 4 = 0$ (c) $x - 2y + 3 = 0$

5. Write down the equation of the line passing through $(2, 1)$ and perpendicular to

    (a) $3x + y - 2 = 0$ (b) $2x - 4y - 1 = 0$

    Draw a sketch showing all four lines on the same set of axes.

6. Write down the equation of the line passing through $(3, -2)$ and parallel to

    (a) $5x - y + 3 = 0$ (b) $x + 7y - 5 = 0$

7. $A(1, 5)$ and $B(4, 9)$ are two adjacent vertices of a square. Find the equation of the line on which the side BC of the square lies. How long are the sides of this square?

Formulae for Finding the Equation of a Line

Straight lines play a major role in graphical analysis and it is important to be able to find their equations easily. This section gives two formulae derived from the commonest ways in which a straight line is defined.

The appropriate formula can then be used to write down the equation of a particular line.

A line with gradient m and passing through the point (x_1, y_1)

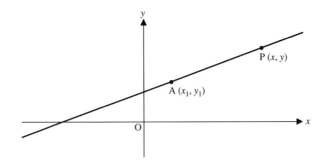

$P(x, y)$ is a point on the line if and only if the gradient of AP is m

i.e. $\dfrac{y - y_1}{x - x_1} = m$

$\Rightarrow$ $\boldsymbol{y - y_1 = m(x - x_1)}$ [1]

A line passing through (x_1, y_1) and (x_2, y_2)

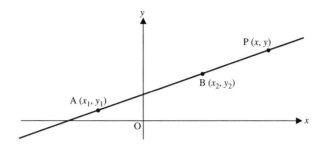

The gradient of AB is $\dfrac{y_2 - y_1}{x_2 - x_1}$

so the formula given in [1] becomes

$$y - y_1 = \left[\frac{y_2 - y_1}{x_2 - x_1}\right](x - x_1) \qquad\qquad [2]$$

Examples 7c

1. Find the equation of the line with gradient $-\frac{1}{3}$ and passing through $(2, -1)$

Using [1] with $m = -\frac{1}{3}$, $x_1 = 2$ and $y_1 = -1$ gives

$$y - (-1) = -\tfrac{1}{3}(x - 2)$$

$\Rightarrow \qquad x + 3y + 1 = 0$

Alternatively the equation of this line can be found from the standard form of the equation of a straight line, i.e. $y = mx + c$

Using $y = mx + c$ and $m = -\frac{1}{3}$ we have

$$y = -\tfrac{1}{3}x + c$$

The point $(2, -1)$ lies on this line so its coordinates satisfy the equation, i.e.

$$-1 = -\tfrac{1}{3}(2) + c \quad \Rightarrow \quad c = -\tfrac{1}{3}$$

Therefore $\qquad y = -\tfrac{1}{3}x - \tfrac{1}{3}$

$\Rightarrow \qquad x + 3y + 1 = 0$

2. Find the equation of the line through the points $(1, -2)$, $(3, 5)$

Using formula [2] with $x_1 = 1$, $y_1 = -2$, $x_2 = 3$ and $y_2 = 5$ gives

$$y - (-2) = \frac{-2 - 5}{1 - 3}(x - 1)$$

$\Rightarrow \qquad 7x - 2y - 11 = 0$

The worked examples in this book necessarily contain a lot of explanation but this should not mislead readers into thinking that their solutions must be equally long. The temptation to 'overwork' a problem should be avoided, particularly in the case of coordinate geometry problems which are basically simple. With practice, any of the methods illustrated above enable the equation of a straight line to be written down directly.

3. Find the equation of the line through $(1, 2)$ which is perpendicular to the line $3x - 7y + 2 = 0$

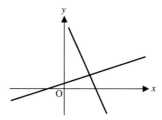

Expressing $3x - 7y + 2 = 0$ in standard form gives $y = \frac{3}{7}x + \frac{2}{7}$

Hence the given line has gradient $\frac{3}{7}$.

So the required line has a gradient of $-\frac{7}{3}$ and it passes through $(1, 2)$

Using $y - y_1 = m(x - x_1)$ gives its equation as

$$y - 2 = -\frac{7}{3}(x - 1) \quad \Rightarrow \quad 7x + 3y - 13 = 0$$

In the last example note that the line perpendicular to

$$3x - 7y + 2 = 0$$

has equation $\qquad 7x + 3y - 13 = 0$

i.e. the coefficients of x and y have been transposed and the sign between the x and y terms has changed. This is a particular example of the general fact that

given a line with equation $ax + by + c = 0$ then the equation of any perpendicular line is $bx - ay + k = 0$

This property of perpendicular lines can be used to shorten the working of problems, e.g. to find the equation of the line passing through $(2, -6)$ which is perpendicular to the line $5x - y + 3 = 0$, we can say that the required line has an equation of the form $y + 5x + k = 0$ and then use the fact that the coordinates $(2, -6)$ satisfy this equation to find the value of k.

EXERCISE 7c

1. Find the equation of the line with the given gradient and passing through the given point.
   (a) 3, (4, 9)
   (b) -5, (2, -4)
   (c) $\frac{1}{4}$, (4, 0)
   (d) 0, (-1, 5)
   (e) $-\frac{2}{5}$, $(\frac{1}{2}, 4)$
   (f) $-\frac{3}{8}$, $(\frac{22}{5}, -\frac{5}{2})$

2. Find the equation of the line passing through the points
   (a) (0, 1), (2, 4)
   (b) (-1, 2), (1, 5)
   (c) (3, -1), (3, 2)
   (d) (0, 0), (2, 5)
   (e) (3, -1), (-5, 2)
   (f) (-5, -1), (0, -7)

3. Determine which of the following pairs of lines are perpendicular.
   (a) $x - 2y + 4 = 0$ and $2x + y - 3 = 0$
   (b) $x + 3y - 6 = 0$ and $3x + y + 2 = 0$
   (c) $x + 3y - 2 = 0$ and $y = 3x + 2$
   (d) $y + 2x + 1 = 0$ and $x = 2y - 4$

4. Find the equation of the line through the point $(5, 2)$ and perpendicular to the line $x - y + 2 = 0$

5. Find the equation of the line with gradient 3 and y-intercept 5.

6. Find the equation of the line with gradient -1 and x-intercept -4.

7. Find the equation of the line through the origin which is parallel to the line $4x + 2y - 5 = 0$

8. The line $4x - 5y + 20 = 0$ cuts the x-axis at A and the y-axis at B. Find the coordinates of the vertices of $\triangle$OAB.

9. Find the equation of the altitude through O of the triangle OAB defined in Question 8.

10. Find the equation of the perpendicular from $(5, 3)$ to the line $2x - y + 4 = 0$

11. The points A$(1, 4)$ and B$(5, 7)$ are two adjacent vertices of a parallelogram ABCD. The point C$(7, 10)$ is another vertex of the parallelogram. Find the equation of the side CD.

INTERSECTION

The point where two lines (or curves) cut is called a point of intersection.

The coordinates of the point (s) of intersection can be found by solving the equations simultaneously.

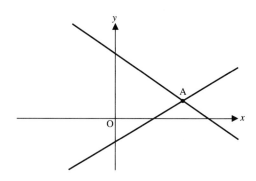

If A is the point of intersection of the lines $y - 3x + 1 = 0$ [1]

and $y + x - 2 = 0$ [2]

then the coordinates of A satisfy both of these equations. A can be found by solving [1] and [2] simultaneously, i.e.

$[2] - [1] \quad \Rightarrow \quad 4x - 3 = 0 \quad \Rightarrow \quad x = \frac{3}{4}$ and $y = \frac{5}{4}$

Therefore $(\frac{3}{4}, \frac{5}{4})$ is the point of intersection.

Note that the coordinates of A can also be found using a graphics calculator or a graph drawing package on a computer.

PROBLEMS

Example 7d

A circle has radius 4 and its centre is the point $C(5, 3)$.

Show that the points $A(5, -1)$ and $B(1, 3)$ are on the circumference of the circle.

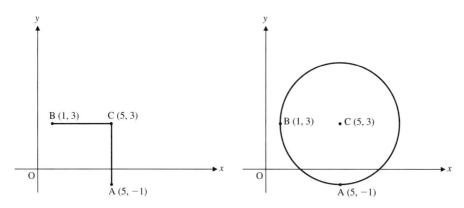

From the diagram, $BC = 4$

The radius is 4,

$\therefore$ B is on the circumference.

Similarly $AC = 4$,

$\therefore$ A is on the circumference.

Note how the diagram helped to give a simple and direct solution. It is always a good idea to draw a sketch; a fairly accurate one can be drawn using squared paper.

MIXED EXERCISE 7

1. Find the coordinates of the point of intersection of each pair of lines.
   (a) $y = 3x - 5,\ y = 4 - 5x$
   (b) $2x + y - 3 = 0,\ x - 3y + 4 = 0$
   (c) $y = 2x - 9,\ 4x - 7y + 5 = 0$

2. Show that the triangle whose vertices are $(1, 1)$, $(3, 2)$ and $(2, -1)$ is isosceles.

3. Find the area of the triangular region enclosed by the x- and y-axes and the line $2x - y - 1 = 0$

4. Find the coordinates of the triangular region enclosed by the lines $y = 0$, $y = x + 5$ and $x + 2y - 6 = 0$

5. Write down the equation of the line that passes through the point $(2, -3)$ and has a gradient of -2.

6. Find the equation of the line through $A(5, 2)$ which is perpendicular to the line $y = 3x - 5$. Hence find the coordinates of the foot of the perpendicular from A to the line.

7. Write down the equation of the line which goes through $(7, 3)$ and which is inclined at $45°$ to the positive direction of the x-axis.

8. Find the equation of the line through the points (a, b) and $(2a, -3b)$

9. The centre of a circle is at the point $C(3, 7)$ and the point $A(5, 3)$ is on the circumference of the circle. Find

   (a)   the radius of the circle,

   (b)   the equation of the line through A that is perpendicular to AC.

10. The equations of two sides of a square are $y = 3x - 1$ and $x + 3y - 6 = 0$. If $(0, -1)$ is one vertex of the square find the coordinates of the other vertices.

11. The lines $y = 2x,\ 2x + y - 12 = 0$ and $y = 2$ enclose a triangular region of the xy-plane. Find

    (a)   the coordinates of the vertices of this region,

    (b)   the area of this region.

12. Find, in terms of a and b, the coordinates of the foot of the perpendicular from the point (a, b) to the line $x + 2y - 4 = 0$

13. The coordinates of a point P are $(t + 1, 2t - 1)$. Sketch the position of P when $t = -1, 0, 1$ and 2 Show that these points are collinear and write down the equation of the line on which they lie.

14. $A(5, 0)$ and $B(0, 8)$ are two vertices of triangle OAB.

    (a)   What is the equation of the bisector of angle AOB.

    (b)   If E is the point of intersection of this bisector and the line through A and B, find the coordinates of E.
         Hence show that $OA : OB = AE : EB$.

*15. Find the equation of the line through the point $(5, 1)$ that is inclined at $60°$ to the positive direction of the x-axis.

*16. The length of a spring was measured and recorded when various different weights were hung on it. The results are shown in the table.

Weight (grams)	50	100	200	500
Length of spring (cm)	0.5	0.85	1.4	3.3

(a) Plot the corresponding values of the length (l cm) against the weight (w grams) and hence explain why it is reasonable to assume that l and w are related by an equation of the form $l = aw + b$.

(b) Draw, by eye, the line that best fits these points and hence find approximate values for the constants a and b.

(c) Interpret the meaning of the constant a in the context of the spring.

(d) Explain why it would not be reasonable to extend this line to estimate what length the spring would be if a weight of 3 kg were hung on it.

CONSOLIDATION B

SUMMARY

Plane Geometry

INTERCEPT THEOREM

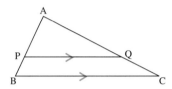

$$PQ \parallel BC \quad \Longleftrightarrow \quad AP : BP = AQ : QC$$

PYTHAGORAS' THEOREM

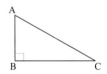

$$\text{In } \triangle ABC, \ \angle B = 90° \quad \Longleftrightarrow \quad AC^2 = AB^2 + BC^2$$

SIMILAR TRIANGLES

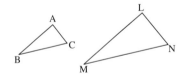

$$\angle A = \angle L, \ \angle B = \angle M, \ \angle C = \angle N$$
$$\Longleftrightarrow \quad AB : LM = BC : MN = AC : LN$$

CONGRUENT TRIANGLES
Two triangles are congruent if, in both triangles,

two sides and the included angle are equal,

or   two angles and a corresponding side are equal,

or   all three sides are equal,

or   two sides are equal and one angle is 90°.

ANGLE BISECTOR THEOREM

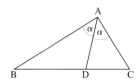

$$AD \text{ bisects } \angle A \quad \Longleftrightarrow \quad BD : DC = AB : AC$$

Trigonometry

In any right-angled triangle

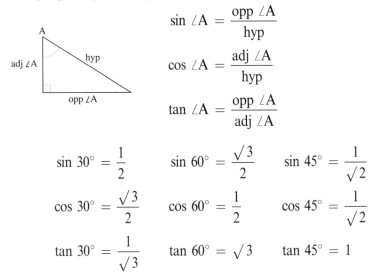

$$\sin \angle A = \frac{\text{opp } \angle A}{\text{hyp}}$$

$$\cos \angle A = \frac{\text{adj } \angle A}{\text{hyp}}$$

$$\tan \angle A = \frac{\text{opp } \angle A}{\text{adj } \angle A}$$

$$\sin 30° = \frac{1}{2} \qquad \sin 60° = \frac{\sqrt{3}}{2} \qquad \sin 45° = \frac{1}{\sqrt{2}}$$

$$\cos 30° = \frac{\sqrt{3}}{2} \qquad \cos 60° = \frac{1}{2} \qquad \cos 45° = \frac{1}{\sqrt{2}}$$

$$\tan 30° = \frac{1}{\sqrt{3}} \qquad \tan 60° = \sqrt{3} \qquad \tan 45° = 1$$

Coordinate Geometry

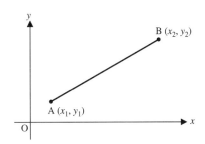

Length of AB is $\sqrt{[(x_2 - x_1)^2 + (y_2 - y_1)^2]}$

Gradient of AB is $\dfrac{y_2 - y_1}{x_2 - x_1}$

Parallel lines have equal gradients.

When two lines are perpendicular, the product of their gradients is -1

The standard equation of a straight line is $y = mx + c$, where m is its gradient and c its intercept on the y-axis.

Any equation of the form $ax + by + c = 0$ is a straight line.

MULTIPLE CHOICE EXERCISE B

TYPE I

1. The equation of the line through $(0, 5)$ with gradient 1 is

   A $y = 5$ **C** $y = x - 5$ **E** $y = 5x + 1$

   B $y = x + 5$ **D** $y = 5x$

2. In $\triangle ABC$, $BC = 3\,cm$, $AC = 4\,cm$ and $AB = 5\,cm$,

   A $A = 90°$ **C** $B = 45°$ **E** $C = 90°$

   B $C = 60°$ **D** $B = 90°$

3.

   A $\tan x = \frac{3}{2}$ **B** $\sin x = \frac{2}{3}$

   C $\cos x = \frac{3}{2}$ **D** $\tan x = \frac{2}{3}$

   E $\sin x = \frac{2}{5}$

4. The length of the line joining $(3, -4)$ to $(-7, 2)$ is

   A $2\sqrt{13}$ **C** $2\sqrt{34}$ **E** 6

   B 16   **D** $2\sqrt{5}$

5. The midpoint of the line joining $(-1, -3)$ to $(3, -5)$ is

   A $(1, 1)$ **C** $(2, -8)$ **E** $(1, -1)$

   B $(0, 0)$ **D** $(1, -4)$

6. The gradient of the line joining $(1, 4)$ and $(-2, 5)$ is

   A $\frac{1}{3}$ **B** $-\frac{1}{3}$ **C** 3   **D** -3 **E** 1.3

7. The gradient of the line perpendicular to the join of $(-1, 5)$ and $(2, -3)$ is

   A $\frac{3}{8}$ **B** $-2\frac{2}{3}$ **C** $\frac{1}{2}$ **D** 2   **E** $2\frac{2}{3}$

8. The equation of the line through the origin and perpendicular to $3x - 2y + 4 = 0$ is

   A $3x + 2y = 0$ **C** $2x + 3y = 0$ **E** $3x - 2y = 0$

   B $2x + 3y + 1 = 0$ **D** $2x - 3y - 1 = 0$

9. In $\triangle ABC$, $AB = 4\,cm$, $\angle A = 90°$, $\angle C = 20°$. $AC =$

   A $4\tan 20°$ **B** $4\tan 70°$ **C** $4\cos 20°$ **D** $4\sin 70°$ **E** $4\sin 20°$

TYPE II

10. A, B and C are the points $(5, 0)$, $(-5, 0)$, $(2, 3)$. AB and BC are perpendicular.

11. The equation of a line is $7x - 2y + 4 = 0$. It has a gradient of $3\frac{1}{2}$

12. The sine of an angle in a triangle is the length of the side opposite the angle.

13. The line joining $(0, 0)$ and $(1, 3)$ is equal in length to the line joining $(0, 1)$ and $(3, 0)$

14. If a line has gradient m and intercept d on the x-axis, its equation is $y = mx - md$

15. The line passing through $(3, 1)$ and $(-2, 5)$ is perpendicular to the line $4y = 5x - 3$

MISCELLANEOUS EXERCISE B

1. Find the equation of the straight line that passes through the points $(3, -1)$ and $(-2, 2)$, giving your answer in the form $ax + by + c = 0$.
   Hence find the coordinates of the point of intersection of the line and the x-axis.
   (UCLES)$_s$

2. The points A and B have coordinates $(8, 7)$ and $(-2, 2)$ respectively. A straight line l passes through A and B and meets the coordinate axes at the points C and D.
   (a) Find, in the form $y = mx + c$, the equation of l.
   (b) Find the length CD, giving your answer in the form $p\sqrt{q}$, where p and q are integers and q is prime.
   (ULEAC)$_s$

3. Find the equation of the straight line which passes through the point $(2, 5)$ and has gradient 3. Find also the coordinates of the point of intersection of this line and the line with equation $x + 2y = 0$
   (OCSEB)

4. A line passes through the point of intersection of the lines with equations $x + 3y - 3 = 0$ and $2x - 3y - 6 = 0$. Find the equation of this line if
   (a) its gradient is $\frac{3}{2}$
   (b) it passes through the point $(0, 3)$

5. The harbour at Swanage is in the direction N 80° E and 32 km from the harbour at Portland. A car ferry is on a bearing of 150° and 20 km from Portland. Find the bearing and distance of the ferry from Swanage.

6. Two of the sides of a parallelogram lie on the lines with equations $y = \frac{1}{2}x$ and $y = 3x$. The vertex that is not on these sides is at the point $(4, 7)$. Find the equations of the lines containing the other two sides of the parallelogram. Hence find the coordinates of the remaining vertices.

7. A vertical wall, 2.7 m high, runs parallel to the wall of a house and is at a horizontal distance of 6.4 m from the house. An extending ladder is placed to rest on the top B of the wall with one end C against the house and the other end A resting on horizontal ground, as shown in the figure.

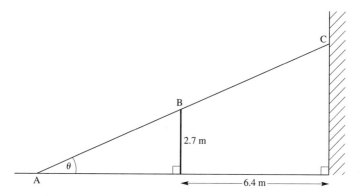

The points A, B and C are in a vertical plane at right angles to the wall and the ladder makes an angle θ, where $0 < \theta < \pi/2$, with the horizontal. Show that the length, y metres, of the ladder is given by

$$y = \frac{2.7}{\sin \theta} + \frac{6.4}{\cos \theta}$$
(ULEAC)$_p$

8. A, B and C are the points $(0, 2)$, $(4, 6)$ and $(10, 0)$ respectively.

(a) Find the lengths AB, BC and CA of the sides of the triangle ABC, and show that $AB^2 + BC^2 = CA^2$.
Deduce the size of angle ABC.

(b) Find the gradients of the lines AB and BC and show how these can be used to confirm your answer in part (i) for the size of angle ABC. (MEI)$_{ps}$

9. The coordinates of the midpoints of the sides of a triangle are $(3, 2)$, $(1, 5)$ and $(-3, -1)$. Find the equations of the lines on which the sides of the triangle lie.

10. The curve $y = 1 + \dfrac{1}{2 + x}$ crosses the x-axis at the point A and the y-axis at the point B.

(a) Calculate the coordinates of A and of B.

(b) Find the equation of the line AB.

(c) Calculate the coordinates of the point of intersection of the line AB and the line with equation $3y = 4x$ (OCSEB)

11. In triangle ABC, $\angle A = 90°$ and the sides AB and AC are each measured as 5.2 cm correct to 1 decimal place. Find the limits between which the size of $\angle C$ lies.

12. The diagram shows a symmetrical roof frame.

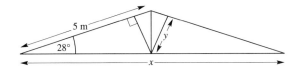

Find the dimensions marked x and y.

13. Find the sizes of the angles of the triangle whose vertices are at the points $(3, 2)$, $(5, 1)$ and $(-3, 2)$

14. The points $P(p, q)$ and $Q(p - 1, 2q + 2)$ lie on the line $2x - y + 3 = 0$. Find the values of p and q.

15. Prove that for all values of s, the point $(2s, s - \frac{15}{4})$ is equidistant from the points $(4, 2)$ and $(7, -4)$

16. The centre of a circle is the point $(5, 2)$ and the origin is on the circumference.

    (a) Find the equation of the diameter through the origin.

    (b) Find the coordinates of the other end of this diameter.

17. (a) Write down the coordinates of the mid-point M of the line joining $A(0, 1)$ and $B(6, 5)$.

    (b) Show that the line $3x + 2y - 15 = 0$ passes through M and is perpendicular to AB. (ULEAC)

18. The equations of two adjacent sides of a rhombus are $y = 2x + 4$, $y = -\frac{1}{3}x + 4$. If $(12, 0)$ is one vertex and all vertices have positive coordinates, find the coordinates of the other three vertices.

19. In the triangle ABC, D is the foot of the perpendicular from A to BC. Show that $AC \sin C = AB \sin B$

20.

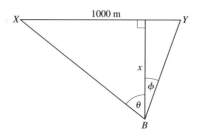

A navigator on a boat at B sights two reference points X and Y which are known to be exactly 1000 m apart, with Y due east of X. The angles θ and ϕ, shown in the diagram, are measured, and the distance x m from B due north to the line XY is calculated from the formula

$$x = \frac{1000}{\tan \theta + + \tan \phi}.$$

Given that the measured values of the angles are $\theta = 52°$ and $\phi = 15°$, each correct to the nearest degree,

(a) find, correct to 1 decimal place, the greatest and least possible values of x,

(b) estimate the greatest possible relative error in taking x to be

$$\frac{1000}{\tan 52° + \tan 15°}$$

(UCLES)$_s$

CHAPTER 8

CIRCLE GEOMETRY

PARTS OF A CIRCLE

We start this chapter with a reminder of the language used to describe parts of a circle.

Part of the circumference is called an *arc*.
If the arc is less than half the circumference it is called a *minor arc*; if it is greater than half the circumference it is called a *major arc*.

A straight line which cuts a circle in two distinct points is called a *secant*. The part of the line inside the circle is called a *chord*.

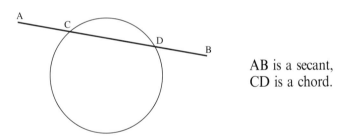

AB is a secant,
CD is a chord.

The area enclosed by two radii and an arc is called a *sector*.

The area enclosed by a chord and an arc is called a *segment*. If the segment is less than half a circle it is called a *minor segment*; if it is greater than half a circle it is called a *major segment*.

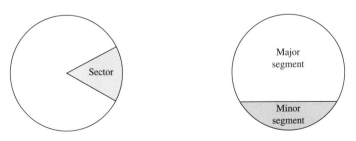

THE ANGLE SUBTENDED BY AN ARC

Consider the points A, B and C
on the circumference of a circle
whose centre is O.

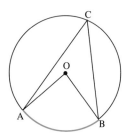

We say that $\angle ACB$ stands on the minor arc AB.

The minor arc AB is said to *subtend* the angle ACB at the circumference (and the
angle is *subtended* by the arc).

In the same way, the arc AB is said to subtend the angle AOB at the centre of the
circle.

Example 8a

A circle of radius 2 units which has its centre at the origin, cuts the *x*-axis at the
points A and B and cuts the *y*-axis at the point C. Prove that $\angle ACB = 90°$

All the information given in the
question, and gleaned from the known
properties of the figure, can be marked
in the diagram as shown. The diagram
can then be referred to as justification
for steps taken in the solution.

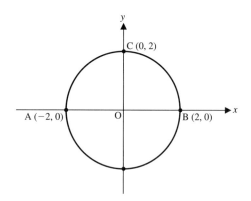

From the diagram, the gradient of AC is $\dfrac{2 - 0}{0 - (-2)} = 1$

and the gradient of BC is $\dfrac{2 - 0}{0 - 2} = -1$

$\therefore$ (gradient of AC) × (gradient of BC) $= -1$

i.e. AC is perpendicular to BC $\Rightarrow$ $\angle ACB = 90°$

1.

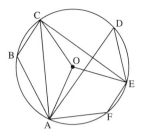

Name the angles subtended

(a) at the circumference by the minor arc AE

(b) at the circumference by the major arc AE

(c) at the centre by the minor arc AC

(d) at the circumference by the major arc AC

(e) at the centre by the minor arc CE

(f) at the circumference by the minor arc CD

(g) at the circumference by the minor arc BC.

2. AB is a chord of a circle, centre O, and M is its midpoint. The radius from O is drawn through M. Prove that OM is perpendicular to AB.

3. C(5, 3) is the centre of a circle of radius 5 units.

(a) Show that this circle cuts the x-axis at A(1, 0) and B(9, 0)

(b) Find the angle subtended at C by the minor arc AB.

(c) The point D is on the major arc AB and DC is perpendicular to AB. Find the coordinates of D and hence find the angle subtended at D by the minor arc AB.

4. A and B are two points on the circumference of a circle centre O. C is a point on the major arc AB. Draw the lines AC, BC, AO, BO and CO, extending the last line to a point D inside the sector AOB. Prove that $\angle$AOD is twice $\angle$ACO and that $\angle$BOD is twice angle $\angle$BCO. Hence show that the angle subtended by the minor arc AB at the centre of the circle is twice the angle that it subtends at the circumference of the circle.

ANGLES IN A CIRCLE

The solutions to questions in the last exercise illustrate two important results.

1) The perpendicular bisector of a chord of a circle goes through the centre of the circle.

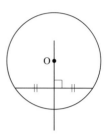

2) The angle subtended by an arc at the centre of a circle is twice the angle subtended at the circumference by the same arc.

Further important results follow from the last fact.

3)

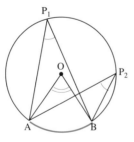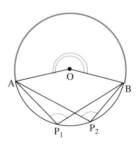

In both diagrams, $\angle AOB = 2\angle P_1 = 2\angle P_2$

So it follows that $\angle P_1 = \angle P_2$

i.e.

  all angles subtended at the circumference by the same arc are equal.

4) A semicircle subtends an angle of 180° at the centre of the circle; therefore it subtends an angle half that size, i.e. 90°, at any point on the circumference. This angle is called the angle in a semicircle.

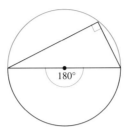

Hence **the angle in a semicircle is 90°**

5) If all four vertices of a quadrilateral ABCD lie on the circumference of a circle, ABCD is called a *cyclic quadrilateral*.

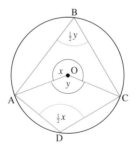

In the diagram, O is the centre of the circle.

∴ $\angle ADC = \frac{1}{2}x$ and $\angle ABC = \frac{1}{2}y$

But $x + y = 360°$, therefore $\angle ADC + \angle ABC = 180°$

i.e. **the opposite angles of a cyclic quadrilateral are supplementary**

EXERCISE 8b

1. Find the size of each marked angle.

(a)

(b)

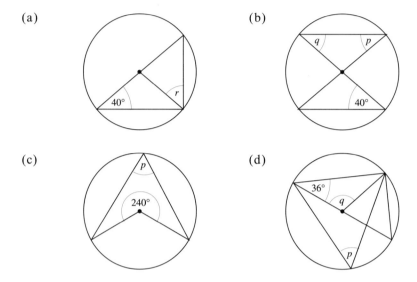

(c)

(d)

2. AB is a diameter of a circle centre O. C is a point on the circumference. D is a point on AC such that OD bisects ∠AOC. Prove that OD is parallel to BC.

3. A triangle has its vertices at the points A(1, 3), B(5, 1) and C(7, 5). Prove that △ABC is right-angled and hence find the coordinates of the centre of the circumcircle of △ABC.

4. Find the size of the angle marked *e* in the diagram.

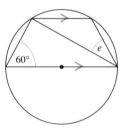

5. AB and CD are two chords of a circle that cut at E. (E is not the centre of the circle.) Show that △s ACE and BDE are similar.

6. A circle with centre O circumscribes an equilateral triangle ABC. The radius drawn through O and the midpoint of AB meets the circumference at D. Prove that △ADO is equilateral.

7. The line joining A(5, 3) and B(4, −2) is a diameter of a circle. If P(a, b) is a point on the circumference find a relationship between a and b.

8. ABCD is a cyclic quadrilateral. The side CD is produced to a point E outside the circle. Show that ∠ABC = ∠ADE.

9. In the diagram, O is the centre of the circle and CD is perpendicular to AB. If ∠CAB = 30° find the size of each marked angle.

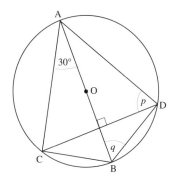

TANGENTS TO CIRCLES

If a line and a circle are drawn on a plane then there are three possibilities for the position of the line in relation to the circle. The line can miss the circle, or it can cut the circle in two distinct points, or it can touch the circle at one point. In the last case the line is called a *tangent* and the point at which it touches the circle is called the *point of contact*.

The length of a tangent drawn from a point to a circle is the distance from that point to the point of contact.

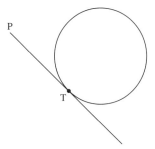

T is the point of contact.

PT is the length of the tangent from P.

Properties of Tangents to Circles

There are two important and useful properties of tangents to circles.

A tangent to a circle is perpendicular to the radius drawn through the point of contact, i.e. AB is perpendicular to OT.

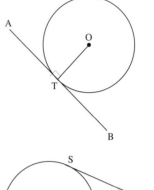

The two tangents drawn from an external point to a circle are equal in length,

i.e. PS = PT

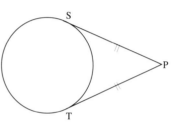

The second property is proved in the following worked example.

Examples 8c

1. PS and PT are two tangents drawn from a point P to a circle whose centre is O. Prove that PT = PS.

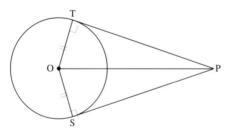

In △s OTP, OSP   ∠T = ∠S = 90°

OS = OT (radii)

and OP is common

∴ △s OTP, OSP are congruent.

Hence PT = PS.

Another useful property follows from the last example, namely

when two tangents are drawn from a point to a circle, the line joining that point to the centre of the circle bisects the angle between the tangents.

2. A circle of radius 10 units is circumscribed by a right-angled isosceles triangle. Find the lengths of the sides of the triangle.

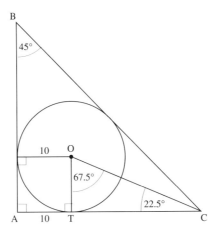

A circle is *circumscribed* by a figure when all the sides of the figure touch the circle. Note also that the circle is inscribed in the figure.

From the diagram

in $\triangle$OTC,   TC $= 10\tan 67.5°$

$= 24.14$

AT $= 10$

$\therefore$ AC $= 34.14 =$ AB

in $\triangle$ABC,   BC $= \sqrt{(34.14^2 + 34.14^2)}$ (Pythagoras)

$= 48.28$

$\therefore$ correct to 3 sf the lengths of the sides of the triangle are 34.1 units, 34.1 units and 48.3 units.

3. The centre of a circle of radius 3 units is the point C(2, 5). The equation of a line, l, is $x + y - 2 = 0$

(a) Find the equation of the line through C, perpendicular to l

(b) Find the distance of C from l and hence determine whether l is a tangent to the circle.

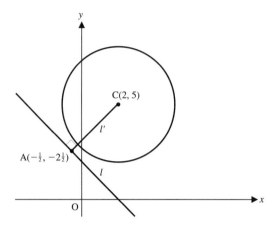

(a)   The line l' is perpendicular to $x + y - 2 = 0$

so its equation is $x - y + k = 0$

The point $(2, 5)$ lies on l'

$\therefore$ $2 - 5 + k = 0$ $\Rightarrow$ $k = 3$

$\therefore$ the equation of l' is $x - y + 3 = 0$

(b)   To find the distance of C from l we need the coordinates of A, the point of intersection of l and l'.

Adding the equations of l and l' gives $2x + 1 = 0$

$\Rightarrow$ $x = -\frac{1}{2}$ and $y = \frac{5}{2}$

so A is the point $(-\frac{1}{2}, \frac{5}{2})$

$\therefore$ $CA = \sqrt{[\{2 - (-\frac{1}{2})\}^2 + \{5 - \frac{5}{2}\}^2]} = 3.54$ to 3 sf

For the line to be a tangent, CA would have to be 3 units exactly (i.e. equal to the radius).

$CA > 3$, therefore l is not a tangent.

EXERCISE 8c

1. The two tangents from a point to a circle of radius 12 units are each of length 20 units. Find the angle between the tangents.

2. Two circles with centres C and O have radii 6 units and 3 units respectively and the distance between O and C is less than 9 units. AB is a tangent to both circles, touching the larger circle at A and the smaller circle at B, where AB is of length 4 units. Find the length of OC.

3. The two tangents from a point A to a circle touch the circle at S and at T. Find the angle between one of the tangents and the chord ST given that the radius of the circle is 5 units and that A is 13 units from the centre of the circle.

4. An equilateral triangle of side 25 cm circumscribes a circle. Find the radius of the circle.

5. AB is a diameter of the circle and C is a point on the circumference. The tangent to the circle at A makes an angle of 30° with the chord AC. Find the angles in $\triangle ABC$.

6. The centre of a circle is at the point $C(4, 8)$ and its radius is 3 units. Find the length of the tangents from the origin to the circle.

7. The points $A(3, 0)$ and $B(13, 0)$ are the ends of a diameter of a circle. The point $C(12, 3)$ is on the circumference. Find the angles in $\triangle ABC$.

8. A circle of radius 6 units has its centre at the point $(9, 0)$. If a tangent from the origin to the circle is inclined to the x-axis at angle α, find $\tan \alpha$.

9. The point $A(6, 8)$ is on the circumference of a circle whose centre is the point $C(3, 5)$. Find the equation of the tangent that touches the circle at A.

10. The line $x - 2y + 4 = 0$ is a tangent to the circle whose centre is the point $C(-1, 2)$.

    (a) Find the equation of the line through C that is perpendicular to the line $x - 2y + 4 = 0$

    (b) Hence find the coordinates of the point of contact of the tangent and the circle.

*11. The equations of the sides of a triangle are $y = 3x$, $y + 3x = 0$ and $3y - x + 12 = 0$. Find the coordinates of the centre of the circumcircle of this triangle.

*12. A, B and C are three points on the circumference of a circle. The tangent to the circle at A makes an angle α with the chord AB. The diameter through A cuts the circle again at D and D is joined to B. Prove that $\angle ACB = \alpha$

*13. A circle touches the y-axis at the origin and goes through the point $A(8, 0)$. C is any point on the circumference. Find the greatest possible area of $\triangle OAC$.

*14. A triangular frame is made to enclose six identical spheres resting on a table and touching each other as shown. Each sphere has a radius of 2 cm. Find the lengths of the sides of the frame.

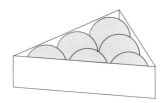

CHAPTER 9

RADIANS, ARCS AND SECTORS

ANGLE UNITS

An angle is a measure of rotation and the units which have been used up to now are the revolution and the degree. It is interesting to note why the number of degrees in a revolution was taken as 360. The ancient Babylonian mathematicians, in the belief that the length of the solar year was 360 days, divided a complete revolution into 360 parts, one for each day as they thought. We now know they did not have the length of the year quite right but the number they used, 360, remains as the number of degrees in one revolution.

Part of an angle smaller than a degree is usually given as a decimal part but until recently the common practice was to divide a degree into 60 minutes ($60'$) and each minute into 60 seconds ($60''$). Limited use is still made of this system.

Now we consider a different unit of rotation which is of great importance in much of the mathematics that follows.

THE RADIAN

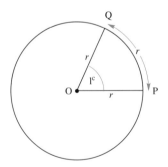

If O is the centre of a circle and an arc PQ is drawn so that its length is equal to the radius of the circle then the angle POQ is called a *radian* (one radian is written 1 rad or 1^c), i.e.

An arc equal in length to the radius of a circle
subtends an angle of 1 radian at the centre.

It follows that the number of radians in a complete revolution is the number of times the radius divides into the circumference.

Now the circumference of a circle is of length $2\pi r$

Therefore the number of radians in a revolution is

$$\frac{2\pi r}{r} = 2\pi$$

i.e. $\qquad$ **2π radians $= 360°$**

Further $\qquad \pi$ radians $= 180°$

and $\qquad \frac{1}{2}\pi$ radians $= 90°$

When an angle is given in terms of π it is usual to omit the radian symbol, i.e. we would write $180° = \pi$ (not $180° = \pi^c$)

If an angle is a simple fraction of $180°$, it is easily expressed in terms of π

e.g. $\qquad 60° = \frac{1}{3}$ of $180° = \frac{1}{3}\pi$

and $\qquad 135° = \frac{3}{4}$ of $180° = \frac{3}{4}\pi$

Conversely, $\qquad \frac{7}{6}\pi = \frac{7}{6}$ of $180° = 210°$

and $\qquad \frac{2}{3}\pi = \frac{2}{3}$ of $180° = 120°$

Angles that are not simple fractions of $180°$, or of π, can be converted by using the relationship $\pi = 180°$ and the value of π from a calculator,

e.g. $\qquad 73° = \dfrac{73}{180} \times \pi = 1.27^c$ (correct to 3 sf)

and $\qquad 2.36^c = \dfrac{2.36}{\pi} \times 180° = 135°$ (correct to the nearest degree)

It helps in visualising the size of a radian to remember that 1 radian is just a little less than $60°$.

$(180° = \pi \text{ rad} \simeq 3.142 \text{ rad} \quad \Rightarrow \quad 1 \text{ rad} = \dfrac{180°}{\pi} \simeq 57°)$

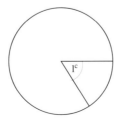

EXERCISE 9a

1. Express each of the following angles in radians as a fraction of π.

   $45°, \quad 150°, \quad 30°, \quad 270°, \quad 225°, \quad 22.5°, \quad 240°, \quad 300°, \quad 315°$

2. Without using a calculator express each of the following angles in degrees.

   $\frac{1}{6}\pi, \quad \pi, \quad \frac{1}{10}\pi, \quad \frac{1}{4}\pi, \quad \frac{5}{6}\pi, \quad \frac{1}{12}\pi, \quad \frac{1}{8}\pi, \quad \frac{4}{3}\pi, \quad \frac{1}{9}\pi, \quad \frac{3}{2}\pi, \quad \frac{4}{9}\pi$

3. Use a calculator to express each of the following angles in radians.

   $35°, \quad 47.2°, \quad 93°, \quad 233°, \quad 14.1°, \quad 117°, \quad 370°$

4. Use a calculator to express each of the following angles in degrees.

   $1.7\,\text{rad}, \quad 3.32\,\text{rad}, \quad 1\,\text{rad}, \quad 2.09\,\text{rad}, \quad 5\,\text{rad}, \quad 6.283\,\text{rad}$

5. Write down, correct to 4 significant figures, the value of

   (a) $\sin 0.2\,\text{rad}$ (b) $\cos 1.1\,\text{rad}$ (c) $\tan 1.5\,\text{rad}$

   (d) $\tan \dfrac{\pi}{4}$ (e) $\cos \dfrac{\pi}{4}$ (f) $\sin \dfrac{\pi}{3}$

MENSURATION OF A CIRCLE

The reader will already be familiar with the formulae for the circumference and the area of a circle, i.e.

$$\text{Circumference} = 2\pi r \quad \text{and} \quad \text{Area} = \pi r^2$$

Now that we have defined a radian, these formulae can be used to derive other results.

The Length of an Arc

Consider an arc which subtends an angle θ at the centre of a circle, *where θ is measured in radians.*

From the definition of a radian, the arc which subtends an angle of 1 radian at the centre of the circle is of length r. Therefore an arc which subtends an angle of θ radians at the centre is of length $r\theta$.

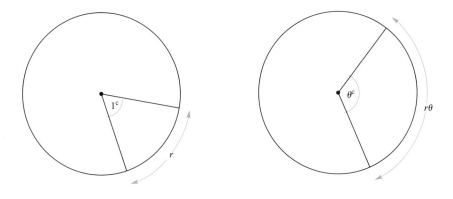

The Area of a Sector

The area of a sector can be thought of as a fraction of the area of the whole circle.

Consider a sector containing an angle of θ radians at the centre of the circle.

The complete angle at the centre of the circle is 2π, hence

$$\frac{\text{Area of sector}}{\text{Area of circle}} = \frac{\theta}{2\pi}$$

$$\Rightarrow \qquad \text{Area of sector} = \frac{\theta}{2\pi} \times \pi r^2$$

$$= \tfrac{1}{2} r^2 \theta$$

We now have two important facts about a circle in which an arc AB subtends an angle θ radians at the centre of the circle.

The length of arc AB $= r\theta$

The area of sector AOB $= \tfrac{1}{2} r^2 \theta$

When solving problems involving arcs and sectors, answers are usually given in terms of π. If a numerical answer is required it will be asked for specifically.

Examples 9b

1. An elastic belt is placed round the rim of a pulley of radius 5 cm. One point on the belt is pulled directly away from the centre, P, of the pulley, until it is at A, 10 cm from P. Find the length of the belt that is in contact with the rim of the pulley.

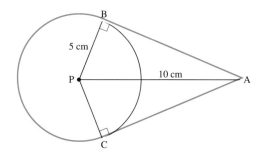

The belt leaves the pulley at B and at C. At these two points the belt is a tangent to the rim, so AB is perpendicular to the radius BP. Similarly, AC and PC are perpendicular.

In $\triangle$ABP, BP $= 5$ cm, AP $= 10$ cm and $\angle$ABP $= 90°$

Therefore $\quad\quad \cos APB = \frac{5}{10} = \frac{1}{2}$

$\Rightarrow \quad\quad\quad\quad\quad \angle APB = 60° = \frac{1}{3}\pi$

Similarly $\quad\quad\quad \angle APC = \frac{1}{3}\pi$

The angle subtended at P by the major arc BC is given by

$$2\pi - \angle BPA - \angle CPA = 2\pi - \tfrac{1}{3}\pi - \tfrac{1}{3}\pi = \tfrac{4}{3}\pi$$

The length of an arc is given by $r\theta$,

therefore the length of the major arc BC is $5 \times \frac{4}{3}\pi = \frac{20}{3}\pi$

i.e. the length of belt in contact with the pulley is $\frac{20}{3}\pi$ cm.

2. AB is a chord of a circle with centre O and radius 4 cm. AB is of length 4 cm and divides the circle into two segments. Find, correct to two decimal places, the area of the minor segment.

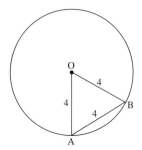

Each side of $\triangle$AOB is 4 cm,

$\therefore$ AOB is an equilateral triangle

$\therefore$ each angle is 60°,

Area of sector $AOB = \frac{1}{2}r^2\theta$

$$= \frac{1}{2}(4^2)(\frac{1}{3}\pi)$$

$$= \frac{8}{3}\pi$$

Area of minor segment $=$ area of sector AOB $-$ area of $\triangle AOB$

$$= \frac{8}{3}\pi - \frac{1}{2}(4)(4)(\sin 60°)$$

$$= 8.378 - 6.928$$

$$= 1.450$$

The area of the minor segment is $1.45\,\text{cm}^2$ correct to 2 dp.

EXERCISE 9b

In questions 1 to 10, s is the length of an arc subtending an angle θ at the centre of a circle of radius r, and a is the area of the corresponding sector. Complete the table.

	$s\,(\text{cm})$	θ	$r\,(\text{cm})$	$a\,(\text{cm}^2)$
1.		$30°$	4	
2.		$\frac{5}{6}\pi$	10	
3.	15	π		
4.	20	$\frac{4}{5}\pi$		
5.		$135°$	8	
6.			2	π
7.			5	12
8.		$\frac{1}{6}\pi$		3π
9.	5π			20π

10. Calculate, in degrees, the angle subtended at the centre of a circle of radius 2.7 cm by an arc of length 6.9 cm.

11. Calculate, in radians, the angle at the centre of a circle of radius 83 mm contained in a sector of area $974\,\text{mm}^2$.

12. The diameter of the moon is about 3445 km and the distance between the moon and earth is about 382 100 km. Find the angle subtended at a point on the earth's surface by the moon (give your answer as a decimal part of a degree to 2 dp).

13. In a circle with centre O and radius 5 cm, AB is a chord of length 8 cm. Find

    (a)   the area of triangle AOB

    (b)   the area of the sector AOB (in square centimetres, correct to 3 sf).

14. A chord of length 10 mm divides a circle of radius 7 mm into two segments. Find the area of each segment.

15. A chord PQ, of length 12.6 cm, subtends an angle of $\frac{2}{3}\pi$ at the centre of a circle. Find

    (a)   the length of the arc PQ

    (b)   the area of the minor segment cut off by the chord PQ.

16. A curve in the track of a railway line is a circular arc of length 400 m and radius 1200 m. Through what angle does the direction of the track turn?

17. Two discs, of radii 5 cm and 12 cm, are placed, partly overlapping, on a table. If their centres are 13 cm apart find the perimeter of the 'figure-eight' shape.

18. Two circles, each of radius 14 cm, are drawn with their centres 20 cm apart. Find the length of their common chord. Find also the area common to the two circles.

*19. A chord of a circle subtends an angle of θ radians at the centre of the circle. The area of the minor segment cut off by the chord is one eighth of the area of the circle.
     Prove that $4\theta = \pi + 4\sin\theta$

*20. A chord PQ of length $6a$ is drawn in a circle of radius $10a$. The tangents to the circle at P and Q meet at R. Find the area enclosed by PR, QR and the minor arc PQ.

*21. Two discs are placed, in contact with each other, on a table. Their radii are 4 cm and 9 cm. An elastic band is stretched round the pair of discs. Calculate

    (a)   the angle subtended at the centre of the smaller disc by the arc that is in contact with the elastic band.

    (b)   the length of the part of the band that is in contact with the smaller disc.

    (c)   the length of the part of the band that is in contact with the larger disc.

    (d)   the total length of the stretched band.

CHAPTER 10

FUNCTIONS

MAPPINGS

If, on a calculator, the number 2 is entered and then the x^2 button is pressed, the display shows the number 4

We say that 2 is mapped to 2^2 or $2 \rightarrow 2^2$

Under the same rule, i.e. squaring the input number,
$3 \rightarrow 9$, $25 \rightarrow 625$, $0.5 \rightarrow 0.25$, $-4 \rightarrow 16$ and in fact,

(any real number) $\rightarrow$ (the square of that number)

The last statement can be expressed more briefly as

$x \rightarrow x^2$

where x is any real number.

This mapping can be represented graphically by plotting values of x^2 against values of x

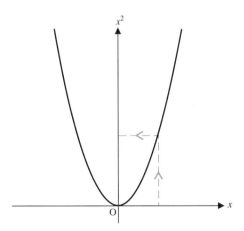

The graph, and knowledge of what happens when we square a number, show that one input number gives just one output number.

Now consider the mapping which maps a number to its square roots; the rule by which, for example, $4 \to +2$ and -2

This rule gives a real output only if the input number is greater than zero (negative numbers do not have real square roots). This mapping can now be written in general terms as

$$x \to \pm\sqrt{x} \quad \text{for} \quad x \geqslant 0$$

The graphical representation of this mapping is shown below.

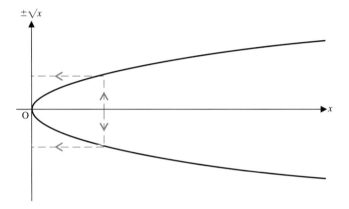

This time we notice that one input value gives two output values.

From these two examples, we can see that a mapping is a rule for changing a number to another number or numbers.

FUNCTIONS

Under the first mapping, $x \to x^2$, one input number gives one output number. However, for the second mapping, $x \to \pm\sqrt{x}$, one input number gives two output numbers.

We use the word *function* for any rule that gives the same kind of result as the first mapping, i.e. one input value gives one output value.

A function is a rule which maps a single number to another single number.

The second mapping does not satisfy this condition so we cannot call it a function.

Consider again what we can now call the function for which $x \to x^2$
Using f for 'function' and the symbol : to mean 'such that', we can write
$f : x \to x^2$

We use the notation $f(x)$ to represent the output values of the function.

e.g.

for $f : x \to x^2$, we have $f(x) = x^2$

Examples 10a

1. Determine whether these mappings are functions,

(a) $x \to \dfrac{1}{x}$ (b) $x \to y$ where $y^2 - x = 0$

(a) For any value of x, except $x = 0$, $\dfrac{1}{x}$ has a single value, therefore $x \to \dfrac{1}{x}$ is a function provided that $x = 0$ is excluded.

Note that $\dfrac{1}{0}$ is meaningless, so to make this mapping a function we have to exclude 0 as an input value.

The function can be described by $f(x) = \dfrac{1}{x}$, $x \neq 0$

(b) If, for example, we input $x = 4$, then the output is the value of y when

$$y^2 - 4 = 0 \quad \Rightarrow \quad y = 2 \text{ and } y = -2$$

therefore an input gives more than one value for the output,
so $x \to y$ where $y^2 - x = 0$ is not a function.

2. If $f(x) = 2x^2 - 5$, find $f(3)$ and $f(-1)$

As $f(x)$ is the output of the mapping, $f(3)$ is the output when 3 is the input, i.e. $f(3)$ is the value of $2x^2 - 5$ when $x = 3$

$$f(3) = 2(3)^2 - 5 = 13$$
$$f(-1) = 2(-1)^2 - 5 = -3$$

EXERCISE 10a

1. Determine which of these mappings are functions.

   (a) $x \to 2x - 1$ (b) $x \to x^3 + 3$ (c) $x \to \dfrac{1}{x-1}$

   (d) $x \to k$ where $k^2 = x$ (e) $x \to \sqrt{x}$

   (f) $x \to$ the length of the line from the origin to $(0, x)$

   (g) $x \to$ the greatest integer less than or equal to x

   (h) $x \to$ the height of a triangle whose area is x

2. If $f(x) = 5x - 4$ find $f(0)$, $f(-4)$

3. If $f(x) = 3x^2 + 25$ find $f(0)$, $f(8)$

4. If $f(x) =$ the value of x correct to the nearest integer,
   find $f(1.25)$, $f(-3.5)$, $f(12.49)$

5. If $f(x) = \sin x$, find $f(\tfrac{1}{2}\pi)$, $f(\tfrac{2}{3}\pi)$

DOMAIN AND RANGE

We have assumed that we can use any real number as an input for a function unless some particular numbers have to be excluded because they do not give real numbers as output.

The set of inputs for a function is called the *domain* of the function.

The domain does not have to contain all possible inputs; it can be as wide, or as restricted, as we choose to make it. Hence to define a function fully, the domain must be stated.

If the domain is not stated, we assume that it is the set of all real numbers ($\mathbb{R}$).

Consider the mapping $x \to x^2 + 3$

We can define a function f for this mapping over any domain we choose. Some examples, together with their graphs are given overleaf.

● $f(x) = x^2 + 3$ for $x \in \mathbb{R}$

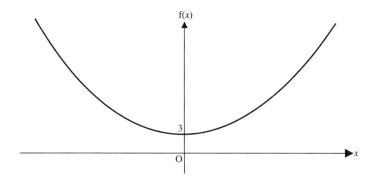

● $f(x) = x^2 + 3$ for $x \geqslant 0$

Note that the point on the curve where $x = 0$ is included and we denote this on the curve by a solid circle.

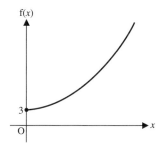

If the domain were $x > 0$, then the point would not be part of the curve and we indicate this fact by using an open circle.

● $f(x) = x^2 + 3$ for $x = 1, 2, 3, 4, 5$

This time the graphical representation consists of just five discrete (i.e. separate) points.

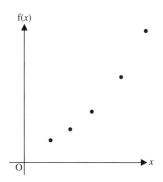

For each domain, there is a corresponding set of output numbers.

The set of output numbers is called the *range* or *image-set* of the function.

Thus for the first function defined, the range is $f(x) \geqslant 3$ and for the second function, the range is also $f(x) \geqslant 3$. For the third function, the range is the set $\{4, 7, 12, 19, 28\}$.

Sometimes a function can be made up from more than one mapping, where each mapping is defined for a different domain. This is illustrated in the next worked example.

Example 10b

The function, f, is defined by $f(x) = x^2$ for $x \leqslant 0$

and $f(x) = x$ for $x > 0$

(a)   Find $f(4)$ and $f(-4)$ (b)   Sketch the graph of f.

(c)   Give the range of f.

(a)   For $x > 0$, $f(x) = x$, $\therefore$ $f(4) = 4$

For $x \leqslant 0$, $f(x) = x^2$, $\therefore$ $f(-4) = (-4)^2 = 16$

(b)   To sketch the graph of a function, we can apply our knowledge of lines and curves in the xy-plane to the equation $y = f(x)$. In this way we can interpret $f(x) = x$ for $x > 0$, as that part of the line $y = x$ which corresponds to positive values of x

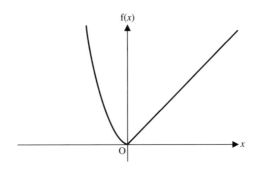

(c)   The range of f is $f(x) \geqslant 0$

1. Find the range of each of the following functions.

   (a) $f(x) = 2x - 3$ for $x \geqslant 0$

   (b) $f(x) = x^2 - 5$ for $x \leqslant 0$

   (c) $f(x) = 1 - x$ for $x \leqslant 1$

   (d) $f(x) = 1/x$ for $x \geqslant 2$

2. Draw a sketch graph of each function given in Question 1.

3. The function f is such that $f(x) = -x$ for $x < 0$
   and $f(x) = x$ for $x \geqslant 0$

   (a) Find the value of $f(5)$, $f(-4)$, $f(-2)$ and $f(0)$

   (b) Sketch the graph of the function.

4. The function f is such that $f(x) = x$ for $0 \leqslant x \leqslant 5$
   and $f(x) = 5$ for $x > 5$

   (a) Find the value of $f(0)$, $f(2)$, $f(4)$, $f(5)$ and $f(7)$

   (b) Sketch the graph of the function.

   (c) Give the range of the function.

5. In Utopia, the tax on earned income is calculated as follows. The first £20 000 is tax free and remaining income is taxed at 20%.

   (a) Find the tax payable on an earned income of £15 000 and of £45 000

   (b) Taking x as the number of pounds of earned income and y as the number of pounds of tax payable, define a function f such that $y = f(x)$. Draw a sketch of the function and state the domain and range.

CURVE SKETCHING

When functions have similar definitions they usually have common properties and graphs of the same form. If the common characteristics of a group of functions are known, the graph of any one particular member of the group can be sketched without having to plot points.

QUADRATIC FUNCTIONS

The general form of a quadratic function is

$$\mathbf{f}(x) = ax^2 + bx + c \ \text{ for } \ x \in \mathbb{R}$$

where a, b and c are constants and $a \neq 0$

When a graphics calculator, or a computer, is used to draw the graphs of quadratic functions for a variety of values of a, b and c, the basic shape of the curve is always the same. This shape is called a *parabola*.

Every parabola has an axis of symmetry which goes through the vertex, i.e. the point where the curve turns back upon itself.

If the coefficient of x^2 is positive, i.e. $a > 0$,
then $f(x)$ has a least value,
and the parabola looks like this.

If the coefficient of x^2 is negative, i.e. $a < 0$,
then $f(x)$ has a greatest value
and the curve is this way up.

These properties of the graph of a quadratic function can be proved algebraically.

For $f(x) = ax^2 + bx + c$, 'completing the square' on the RHS and simplifying, gives

$$f(x) = \left[\frac{4ac - b^2}{4a}\right] + a\left[x + \frac{b}{2a}\right]^2 \qquad [1]$$

Now the first bracket is constant and, as the second bracket is squared, its value is zero when $x = -\dfrac{b}{2a}$ and greater than zero for all other values of x.

Hence

when a is positive,

$$f(x) = ax^2 + bx + c \text{ has a least value when } x = -\frac{b}{2a}$$

and when a is negative,

$$f(x) = ax^2 + bx + c \text{ has a greatest value when } x = -\frac{b}{2a}$$

Further, taking values of x that are symmetrical about $x = -\dfrac{b}{2a}$,

e.g. $x = \pm k - \dfrac{b}{2a}$, we see from [1] that

$$f\left(k - \frac{b}{2a}\right) = f\left(-k - \frac{b}{2a}\right) = \left[\frac{4ac - b^2}{4a}\right] + ak^2$$

i.e. **the value of $f(x)$ is symmetrical about $x = -\dfrac{b}{2a}$**

These properties can now be used to draw *sketches* of the graphs of quadratic functions.

Examples 10c

1. **Find the greatest or least value of the function given by $f(x) = 2x^2 - 7x - 4$ and hence sketch the graph of $f(x)$.**

$f(x) = 2x^2 - 7x - 4 \quad \Rightarrow \quad a = 2, \ b = -7 \text{ and } c = -4$

As $a > 0$, $f(x)$ has a least value

and this occurs when $x = -\dfrac{b}{2a} = \dfrac{7}{4}$

$\therefore$ the least value of $f(x)$ is $f(\tfrac{7}{4}) = 2(\tfrac{7}{4})^2 - 7(\tfrac{7}{4}) - 4$

$$= -\tfrac{81}{8}$$

We now have one point on the graph of $f(x)$ and we know that the curve is symmetrical about this value of x. However, to locate the curve more accurately we need another point; we use $f(0)$ as it is easy to find.

$f(0) = -4$

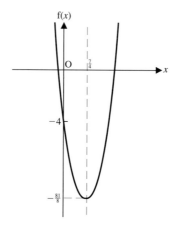

2. **Draw a quick sketch of the graph of** $f(x) = (1 - 2x)(x + 3)$

The coefficient of x^2 is negative, so $f(x)$ has a greatest value.
The curve cuts the x-axis when $f(x) = 0$

When $f(x) = 0$, $\qquad (1 - 2x)(x + 3) = 0$

$\Rightarrow \qquad x = \frac{1}{2}$ or -3

The average of these values is $-\frac{5}{4}$, so the curve is symmetrical about $x = -\frac{5}{4}$

We now have enough information to draw a quick sketch, but note that this method is suitable only when the quadratic function factorises.

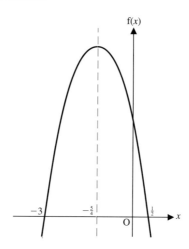

EXERCISE 10c

1. Find the greatest or least value of $f(x)$ where $f(x)$ is

   (a) $x^2 - 3x + 5$ (b) $2x^2 - 4x + 5$ (c) $3 - 2x - x^2$

2. Find the range of f where $f(x)$ is

   (a) $7 + x - x^2$ (b) $x^2 - 2$ (c) $2x - x^2$

3. Sketch the graph of each of the following quadratic functions, showing the greatest or least value and the value of x at which it occurs.

   (a) $x^2 - 2x + 5$ (b) $x^2 + 4x - 8$ (c) $2x^2 - 6x + 3$
   (d) $4 - 7x - x^2$ (e) $x^2 - 10$ (f) $2 - 5x - 3x^2$

4. Draw a quick sketch of each of the following functions.

   (a) $(x - 1)(x - 3)$ (b) $(x + 2)(x - 4)$ (c) $(2x - 1)(x - 3)$
   (d) $(1 + x)(2 - x)$ (e) $x^2 - 9$ (f) $3x^2$

CUBIC FUNCTIONS

The general form of a cubic function is
$$f(x) = ax^3 + bx^2 + cx + d$$
where a, b, c and d, are constants and $a \neq 0$

Investigating the curve $y = ax^3 + bx^2 + cx + d$ for a variety of values of a, b, c and d shows that the shape of the curve is

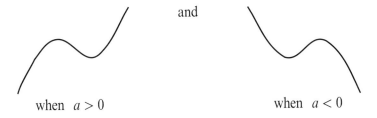

and

when $a > 0$ when $a < 0$

Sometimes there are no turning points and the curve looks like this

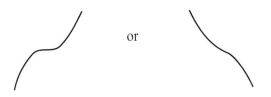

or

THE FUNCTION $f(x) = \dfrac{1}{x}$

Now consider the familiar function $f(x) = 1/x$ and its graph. From its form we can infer various properties of $f(x)$.

1) As the value of x increases, the value of $f(x)$ gets closer to zero,
e.g. when $x = 100$, $f(x) = 1/100$
and when $x = 1000$, $f(x) = 1/1000$
We write this as $x \to \infty$, $f(x) \to 0$

Similarly as the value of x decreases, i.e. as $x \to -\infty$, the value of $f(x)$ again gets closer to zero, i.e. $f(x) \to 0$

2) $f(x)$ does not exist when $x = 0$, so this value of x must be excluded from the domain of f,
x can get as close as we like to zero however, and can approach zero in two ways.

If $x \to 0$ from above (i.e. from positive values, $- - | \longleftarrow - -)$
then $f(x) \to \infty$ $\scriptstyle()$

If $x \to 0$ from below (i.e. from negative values, $- - \longrightarrow | - -)$
then $f(x) \to -\infty$ $\scriptstyle()$

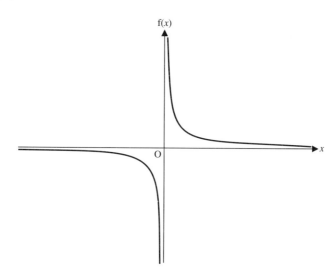

Notice that, as $x \to \pm\infty$, the curve gets closer to the x-axis but does not cross it. Also, as $x \to 0$, the curve approaches the y-axis but again does not cross it. We say that the x-axis and the y-axis are *asymptotes* to the curve.

EXERCISE 10d

1. Find the values of x where the curve $y = f(x)$ cuts the x-axis and sketch the curve when

   (a) $f(x) = x(x-1)(x+1)$ (b) $f(x) = (x^2 - 1)(2 - x)$

2. The basic cubic function is given by $f : x \rightarrow x^3, \ x \in \mathbb{R}$. Draw an accurate graph of this function for values of x from -3 to 3 by plotting points at intervals of 0.25

SIMPLE TRANSFORMATION OF CURVES

Transformations of curves are best appreciated if they can be 'seen', so this section starts with an investigative approach using a graphics calculator or a computer with a graph-drawing package. This exercise is not essential and all the necessary conclusions are drawn analytically in the next part of the text.

EXERCISE 10e

You will need a graphics calculator or computer for this exercise.

1. (a) On the screen, draw the graph of $y = x^3$. Superimpose the graphs of $y = x^3 + 2$ and $y = x^3 - 1$. Clear the screen and again draw the graph of $y = x^3$. This time superimpose the graph of $y = x^3 + c$ for a variety of values of c.

   (b) Describe the transformation that maps the graph of $f(x) = x^3$ to the graph of $g(x) = x^3 + c$

   (c) Repeat (a) and (b) for other simple functions, e.g., x^2, $3x$, $1/x$

2. Use a procedure similar to that described in Question 1 to investigate the relationship between the graphs of $f(x)$ and $f(x + c)$

3. Investigate the relationship between the graphs of

   (a) $f(x)$ and $-f(x)$ (b) $f(x)$ and $f(-x)$

TRANSLATIONS

Consider the function $f(x) = x^3$ and the function $g(x) = x^3 + 2$

For any given value of x, $g(x) = f(x) + 2$. Therefore for equal values of x, points on the curve $y = x^3 + 2$ are 2 units above points on the curve $y = x^3$,

i.e. the curve $y = f(x) + 2$ is a translation of the curve $y = f(x)$ by 2 units in the direction Oy

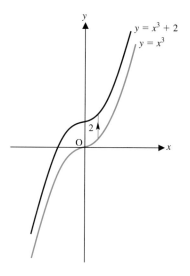

In general, for any function f,
the curve $y = f(x) + c$ is the translation of the curve $y = f(x)$
by c units parallel to the y-axis.

Now consider the curves $y = x^3$ and $y = (x - 2)^3$

The values of y are equal when the value of x in $y = (x - 2)^3$ is 2 units *greater* than the value of x in $y = x^3$, therefore for equal values of y, points on the curve $y = (x - 2)^3$ are 2 units to the *right* of the points on $y = x^3$.

i.e. the curve $y = f(x - 2)$ is a translation of the curve $y = f(x)$ by 2 units in the direction Ox.

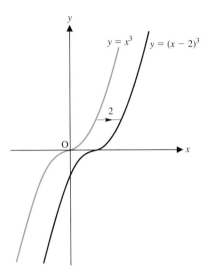

In general, the curve $y = f(x + c)$ is a translation of the curve $y = f(x)$ by c units parallel to the x-axis.

If c is negative the translation is in the direction Ox

If c is positive, the translation is in the direction xO

REFLECTIONS

Consider the function $f(x) = x^2$ and the function $g(x) = -x^2$. For any given value of x, $g(x)$ is equal to $-f(x)$. Therefore for equal values of x, points on the curve $y = -x^2$ are the reflection in the x-axis of points on the curve $y = x^2$

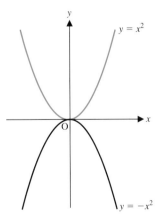

In general, the curve $y = -f(x)$ is the reflection of the curve $y = f(x)$ in the x-axis.

Now consider the function $f(x) = x(x - 1)$ and the function
$g(x) = f(-x)$
Comparing $f(x) = x(x - 1)$ with $g(x) = -x(-x - 1)$ we see that when the inputs to $f(x)$ and to $g(x)$ are equal in value but opposite in sign, $f(x)$ and $g(x)$ have the same value.
Therefore for equal values of y, points on the curve $y = -x(-x - 1)$ are the reflection in the y-axis of points on the curve $y = x(x - 1)$

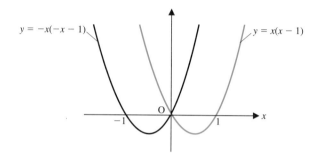

In general, the curve $y = f(-x)$ is the reflection of the curve
$y = f(x)$ in the y-axis.

EVEN FUNCTIONS

A function is even if $f(x) = f(-x)$

Since the curve $y = f(-x)$ is the reflection of the curve $y = f(x)$ in the y-axis, it follows that

when $f(x)$ is an even function,
the curve $y = f(x)$ is symmetrical about Oy

The function $f : x \rightarrow x^2$ is a familiar even function.

$$f(x) = x^2$$

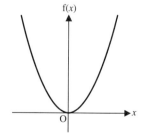

ODD FUNCTIONS

A function is odd if $f(x) = -f(-x)$

As the curve $y = -f(-x)$ is a reflection of the curve $y = f(x)$ in Oy followed by a reflection in Ox, it follows that

when $f(x) = -f(-x)$ the curve $y = f(x)$ has rotational symmetry
of order 2 about the origin.

Some familiar odd functions and their graphs are shown below.

$f(x = x)$ $f(x) = x^3$ $f(x) = \dfrac{1}{x}$

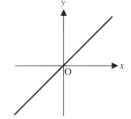 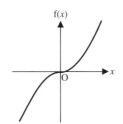 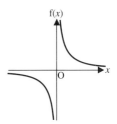

Example 10f

Sketch the curve $y = (x - 1)^3$.

The shape and position of the curve $y = x^3$ is known. If $f(x) = x^3$, then $(x - 1)^3 = f(x - 1)$, so the curve $y = (x - 1)^3$ is a translation of the first curve by one unit in the positive direction of the x-axis.

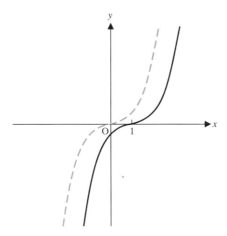

EXERCISE 10f

Sketch each of the following curves, and state whether y is an even or an odd function of x or neither of these.

1. $y = -x^2$

2. $y = -\dfrac{1}{x}$

3. $y = -x^3$

4. $y = 1 + \dfrac{1}{x}$

5. $y = x^2 + 3$

6. $y = \dfrac{1}{x} - 2$

7. $y = (x - 4)^3$

8. $y = x^2 - 9$

9. $y = \dfrac{1}{x - 2}$

10. On the same set of axes sketch the graphs of $f(x) = x^3$, $f(x) = (x + 1)^3$, and $f(x) = -(x + 1)^3$ and $f(x) = 2 - (x + 1)^3$

11. On the same set of axes sketch the lines $y = 2x - 1$ and $y = \frac{1}{2}(x + 1)$. Describe a transformation which maps the first line to the second line.

12. Repeat Question 11 for the curves $y = 1 + \dfrac{1}{x}$ and $y = \dfrac{1}{x - 1}$

13. Find the coordinates of the reflection of the point $(2, 5)$ in the line $y = x$

14. P' is the reflection of the point $P(a, b)$ in the line $y = x$. Find the coordinates of P' in terms of a and b

INVERSE FUNCTIONS

Consider the function f where $f(x) = 2x$ for $x = 2, 3, 4$

Under this function, the domain $\{2, 3, 4\}$ maps to the image-set $\{4, 6, 8\}$ and this is illustrated by the arrow diagram.

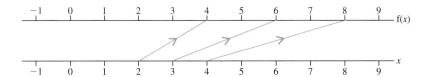

It is possible to reverse this mapping, i.e. we can map each member of the image-set back to the corresponding member of the domain by halving each member of the image-set.

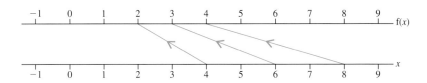

This procedure can be expressed algebraically, i.e.

for $x = 4, 6, 8$, $x \rightarrow \frac{1}{2}x$ maps 4 to 2, 6 to 3 and 8 to 4

This reverse mapping is a function in its own right and it is called the *inverse* function of f where $f(x) = 2x$

Denoting this inverse function by f^{-1} we can write $f^{-1}(x) = \frac{1}{2}x$.

In fact, $f(x) = 2x$ can be reversed for all real values of x and the procedure for doing this is a function.

Therefore, for $f(x) = 2x$, $f^{-1}(x) = \frac{1}{2}x$ is such that f^{-1} reverses f for all real values of x, i.e. f^{-1} maps the output of f to the input of f.

In general, for any function f,

**if there exists a function, g,
that maps the output of f back to its input, i.e. $g : f(x) \rightarrow x$,
then this function is called the inverse of f
and it is denoted by f^{-1}**

THE GRAPH OF A FUNCTION AND ITS INVERSE

Consider the curve that is obtained by reflecting $y = f(x)$ in the line $y = x$. The reflection of a point $A(a, b)$ on the curve $y = f(x)$, is the point A' whose coordinates are (b, a), i.e. interchanging the x and y coordinates of A gives the coordinates of A'.

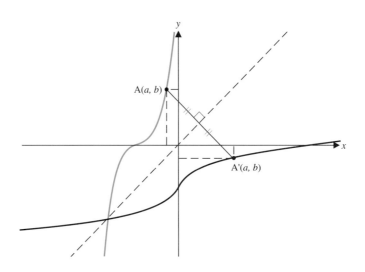

We can therefore obtain the equation of the reflected curve by interchanging x and y in the equation $y = f(x)$

Now the coordinates of A on $y = f(x)$ can be written as $[a, f(a)]$. Therefore the coordinates of A′ on the reflected curve are $[f(a), a]$, i.e. the equation of the reflected curve is such that the output of f is mapped to the input of f.

Hence if the equation of the reflected curve can be written in the form $y = g(x)$, then g is the inverse of f, i.e. $g = f^{-1}$.

To illustrate these properties, consider the curve $y = x^3 + 1$ and its reflection in the line $y = x$

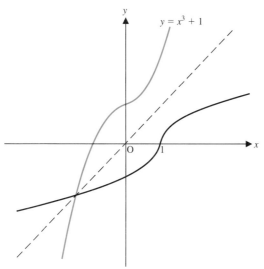

The equation of the reflected curve is given by interchanging x and y in $y = x^3 + 1$ giving $x = y^3 + 1$

We can write this equation in the form $y = \sqrt[3]{(x-1)}$

Therefore for the function $f(x) = x^3 + 1$ the inverse function is given by $f^{-1}(x) = \sqrt[3]{(x-1)}$

Any curve whose equation can be written in the form $y = f(x)$ can be reflected in the line $y = x$. However, this reflected curve may not have an equation that can be written in the form $y = f^{-1}(x)$

Consider the curve $y = x^2$ and its reflection in the line $y = x$

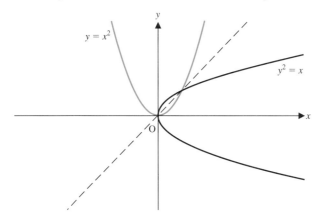

The equation of the image curve is $x = y^2 \implies y = \pm\sqrt{x}$ and $x \to \pm\sqrt{x}$ is not a function.

(We can see this from the diagram as, on the reflected curve, one value of x maps to two values of y. So in this case y cannot be written as a function of x.)

Therefore the function $f : x \to x^2$ does not have an inverse, i.e.

not every function has an inverse.

If we change the definition of f to $f : x \to x^2$ for $x \in \mathbb{R}^+$ then the inverse mapping is

$x \to \sqrt{x}$ for $x \in \mathbb{R}^+$ and this *is* a function, i.e.

$f^{-1}(x) = \sqrt{x}$ for $x \in \mathbb{R}^+$

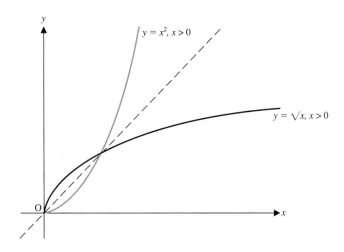

To summarise:

The inverse of a function undoes the function, i.e. it maps the output of a function back to its input.

The inverse of the function f is written f^{-1}

Not all functions have an inverse.

When the curve whose equation is $y = f(x)$ is reflected in the line $y = x$, the equation of the reflected curve is $x = f(y)$
If this equation can be written in the form $y = g(x)$ then g is the inverse of f, i.e. $g(x) = f^{-1}(x)$

Examples 10g

1. Determine whether there is an inverse of the function f given by $f(x) = 2 + \dfrac{1}{x}$

    If f^{-1} exists, express it as a function of x

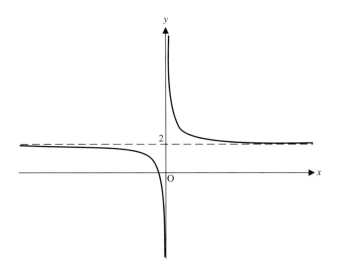

From the sketch of $f(x) = 2 + \dfrac{1}{x}$, we see that one value of $f(x)$ maps to one value of x, therefore the reverse mapping is a function. The equation of the reflection of $y = 2 + \dfrac{1}{x}$ can be written as

$$x = 2 + \frac{1}{y} \quad \Rightarrow \quad y = \frac{1}{x - 2}$$

$\therefore$ when $f(x) = 2 + \dfrac{1}{x}$, $\quad f^{-1}(x) = \dfrac{1}{x - 2}$

2. Find $f^{-1}(4)$ when $f(x) = 5x - 1$

    If $y = f(x)$, i.e. $y = 5x - 1$,

    then $\qquad x = 5y - 1 \quad \Rightarrow \quad y = \frac{1}{5}(x + 1)$

    i.e. $\qquad f^{-1}(x) = \frac{1}{5}(x + 1)$

    $\therefore \qquad f^{-1}(4) = \frac{1}{5}(4 + 1) = 1$

EXERCISE 10g

1. Sketch the graphs of $f(x)$ and $f^{-1}(x)$ on the same axes.

   (a) $f(x) = 3x - 1$ (b) $f(x) = x^3 + 1$

   (c) $f(x) = (x - 1)^3$ (d) $f(x) = 2 - x$

   (e) $f(x) = \dfrac{1}{x - 3}$ (f) $f(x) = \dfrac{1}{x}$

2. Which of the functions given in Question 1 are their own inverses?

3. Determine whether f has an inverse function, and if it does, find it when

   (a) $f(x) = x + 1$ (b) $f(x) = x^2 + 1$

   (c) $f(x) = x^3 + 1$ (d) $f(x) = x^2 - 4, \ x \geqslant 0$

   (e) $f(x) = (x + 1)^4, \ x \geqslant -1$

4. The function f is given by $f(x) = 1 - \dfrac{1}{x}$. Find

   (a) $f^{-1}(4)$ (b) the value of x for which $f^{-1}(x) = 2$

   (c) any values of x for which $f^{-1}(x) = x$

5. If $f(x) = x^3$, find

   (a) $f(2)$ (b) $f^{-1}(27)$ (c) $f^{-1}(1)$

FUNCTION OF A FUNCTION

Consider the two functions f and g where

$$f : x \rightarrow x^2 \qquad \text{and} \qquad g : x \rightarrow \frac{1}{x}$$

These functions can be combined in several ways; they can be added, subtracted, multiplied or divided.

As well as adding, subtracting, multiplying and dividing them, the output of f can be made the input of g, i.e. $x \xrightarrow{f} x^2 \xrightarrow{g} \dfrac{1}{x^2}$ or $g[f(x)] = g(x^2) = \dfrac{1}{x^2}$

Therefore the function $x \to 1/x^2$ is obtained by taking the function g of the function f.

A compound function formed in this way is known as a *function of a function* and it is denoted by gf, or sometimes by g o f,

For example, if $f(x) = x^2$ and $g(x) = 1 - x$ then $gf(x)$ means the function g of $f(x)$,

i.e. $gf(x) = g(x^2) = 1 - x^2$

Similarly $fg(x) = f(1 - x) = (1 - x)^2$

Note that $gf(x)$ is *not* the same as $fg(x)$

EXERCISE 10h

1. If f, g and h are functions defined by $f(x) = x^2$, $g(x) = 1/x$, $h(x) = 1 - x$, find as a function of x

   (a) fg   (b) fh   (c) hg   (d) hf   (e) gf

2. If $f(x) = 2x - 1$ and $g(x) = x^3$ find the value of

   (a) gf(3)   (b) fg(2)   (c) fg(0)   (d) gf(0)

3. Given that $f(x) = 2x$, $g(x) = 1 + x$ and $h(x) = x^2$, find as a function of x

   (a) hg   (b) fhg   (c) ghf

4. The function $f(x) = (2 - x)^2$ can be expressed as a function of a function. Find g and h as functions of x such that $gh(x) = f(x)$

5. Repeat Question 4 when $f(x) = (x + 1)^4$

6. Express the function $f(x)$ as a function g of a function h and define $g(x)$ and $h(x)$, if $f(x)$ is

   (a) $\dfrac{1}{2x + 1}$ (b) $(5x - 6)^4$ (c) $\sqrt{(x^2 - 2)}$

MIXED EXERCISE 10

1. A function f is defined by $f(x) = 1/(1-x)$, $x \neq 1$

   (a) Why is 1 excluded from the domain of f?

   (b) Find the value of $f(-3)$

   (c) Sketch the curve $y = f(x)$

   (d) Find $f^{-1}(x)$ in terms of x and give the domain of f^{-1}

2. Find the greatest or least value of each of the following functions, stating the value of x at which they occur.

   (a) $f(x) = x^2 - 3x + 5$ (b) $f(x) = 2x^2 - 7x + 1$

   (c) $f(x) = (x-1)(x+5)$

3. If $f(x) = x^3$, sketch the following curves on the same set of axes.

   (a) $y = f(x)$ (b) $y = f(x+3)$ (c) $y = f^{-1}(x)$

4. Given that $f(x) = 2x + 1$, $g(x) = x^2$ and $h(x) = 1/x$,

   (a) state whether f, g and h are even or odd functions or neither

   (b) find $fg(2)$, $hg(3)$ and $gf(-1)$

   (c) find, in terms of x, $hfg(x)$ and $gfh(x)$

   (d) if they exist, find $f^{-1}(x)$, $g^{-1}(x)$ and $h^{-1}(x)$

   (e) find the value(s) of x for which $gh(x) = 9$

   (f) does the function $(gh)^{-1}$ exist?

5. Draw sketches of the following curves, showing any asymptotes.

   (a) $y = (x-2)(x-3)(x-4)$ (b) $y = \dfrac{1}{3-x}$ (c) $y = x^2 - 4$

6. The function f is given by $f(x) = \dfrac{1}{2x - 1}$

   (a) find $g(x)$ and $h(x)$ such that $f = gh$

   (b) evaluate $ff(2)$ and $f^{-1}(2)$

7. The functions f, g and h are defined by $f(x) = 2x$, $g(x) = 3x^2$ and $h(x) = x - 1$

   (a) Sketch the curves $y = g(x-3)$ and $y = gf^{-1}(x)$

   (b) Find the value(s) of x for which $f^{-1}(x) = g(x)$

*8. Electronic engineers often need to generate voltages which, as they vary with time, form certain repeating patterns.
One such pattern is called a 'sawtooth' wave, and is such that, when the voltage is plotted against time, the graph looks like this.

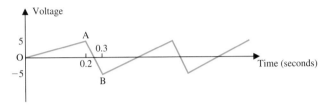

Another such pattern is called a rectangular wave and looks like this.

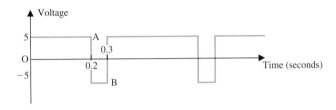

(a) Using Ox for the horizontal axis, describe fully the function that represents the section OAB of
  (i) the sawtooth wave, naming the function f(x),
  (ii) the rectangular wave, naming the function g(x).

(b) The functions f(x) and g(x) can be combined in various ways. For each of the following combinations *sketch* the graph given.
  (i) f(x) + g(x)   (ii) f(x) × g(x)   (iii) fg(x).

CHAPTER 11

INEQUALITIES

MANIPULATING INEQUALITIES

An inequality compares two unequal quantities.

Consider, for example, the two real numbers 3 and 8 for which

$$8 > 3$$

The inequality remains true, i.e. the inequality sign is unchanged, when the same term is added or subtracted on both sides, e.g.

$$8 + 2 > 3 + 2 \quad \Rightarrow \quad 10 > 5$$

and $\quad 8 - 1 > 3 - 1 \quad \Rightarrow \quad 7 > 2$

The inequality sign is unchanged also when both sides are multiplied or divided by a positive quantity, e.g.

$$8 \times 4 > 3 \times 4 \quad \Rightarrow \quad 32 > 12$$

and $\quad 8 \div 2 > 3 \div 2 \quad \Rightarrow \quad 4 > 1\frac{1}{2}$

If, however, both sides are multiplied or divided by a *negative* quantity the inequality is no longer true. For example, if we multiply by -1, the LHS becomes -8 and the RHS becomes -3 so the correct inequality is now LHS < RHS, i.e.

$$8 \times -1 < 3 \times -1 \quad \Rightarrow \quad -8 < -3$$

Similarly, dividing by -2 gives $-4 < -1\frac{1}{2}$

These examples are illustrations of the following general rules.

> **Adding or subtracting a term, or multiplying or dividing both sides by a positive number, does not alter the inequality sign.**

> **Multiplying or dividing both sides by a *negative* number reverses the inequality sign.**

i.e.   **if a, b and k are real numbers, and $a > b$ then,**

$$a + k > b + k \quad \text{for } \textit{all} \text{ values of } k$$
$$ak > bk \qquad \text{for } \textit{positive} \text{ values of } k$$
$$ak < bk \qquad \text{for } \textit{negative} \text{ values of } k$$

SOLVING LINEAR INEQUALITIES

When an inequality contains an unknown quantity, the rules given above can be used to 'solve' it. Whereas the solution of an equation is a value, or values, of the variable, the solution of an inequality is a range, or ranges of values, of the variable.

If the unknown quantity appears only in linear form, we have a *linear inequality* and the solution range has only *one boundary*.

Example 11a

Find the set of values of x that satisfy the inequality $x - 5 < 2x + 1$

$$x - 5 < 2x + 1$$

$\Rightarrow$ $x < 2x + 6$ adding 5 to each side

$\Rightarrow$ $-x < 6$ subtracting $2x$ from each side

$\Rightarrow$ $x > -6$ multiplying both sides by -1

Therefore the set of values of x satisfying the given inequality is

$$x > -6$$

EXERCISE 11a

Solve the following inequalities.

1. $x - 4 < 3 - x$ 2. $x + 3 < 3x - 5$ 3. $x < 4x + 9$

4. $7 - 3x < 13$ 5. $x > 5x - 2$ 6. $2x - 1 < x - 4$

7. $1 - 7x < x + 3$ 8. $2(3x - 5) > 6$ 9. $3(3 - 2x) < 2(3 + x)$

Double Inequalities

Example 11b

Find the range of values of x for which $4 - x < x + 8 < 5 - 2x$

$4 - x < x + 8 < 5 - 2x$ is called a *double inequality* because it contains two inequalities.
i.e. $4 - x < x + 8$ and $x + 8 < 5 - 2x$
We are looking for the set of values of x for which *both* inequalities are satisfied so first we solve them separately.

$$4 - x < x + 8 \qquad\qquad\qquad x + 8 < 5 - 2x$$

$$\Rightarrow \qquad\quad -4 < 2x \qquad\qquad\qquad\quad 3x < -3$$

$$\Rightarrow \qquad\qquad x > -2 \qquad\qquad\qquad\quad x < -1$$

   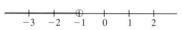

The required set of values must satisfy both of these conditions

i.e.

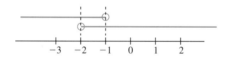

The solution set is $-2 < x < -1$

EXERCISE 11b

For what range(s) of values of x are the following inequalities valid?

1. $x + 2 < 4x - 1 < 2x + 5$

2. $2 - x > 2x + 4 > x$

3. $x - 1 < 3x + 1 < x + 5$

4. $2x - 1 < 3x + 2 < x + 6$

5. $4x - 5 > 5x + 8 > x$

6. $x + 11 > 3x + 1 > 2x + 3$

7. $4 + 5x < 8 + 4x < 6x + 10$

8. $x - 6 > 2 - x > 3x - 18$

SOLVING QUADRATIC INEQUALITIES

A quadratic inequality is one in which the variable appears to the power 2, e.g. $x^2 - 3 > 2x$

The solution is a range or ranges of values of the variable with *two boundaries*.

If the terms in the inequality can be collected and factorised, a graphical solution is easy to find.

Example 11c

Find the range(s) of values of x that satisfy the inequality $x^2 - 3 > 2x$

$$x^2 - 3 > 2x$$

$\Rightarrow \qquad x^2 - 2x - 3 > 0$

$\Rightarrow \qquad (x - 3)(x + 1) > 0$

or $\qquad f(x) > 0$ where $f(x) = (x - 3)(x + 1)$

If we sketch the graph of $f(x)$ then $f(x) > 0$ where the graph is above the x-axis. The values of x corresponding to these portions of the graph satisfy $f(x) > 0$

The points where $f(x) = 0$, i.e. where $x = 3$ and -1 are not part of this solution and this is indicated on the sketch by open circles.

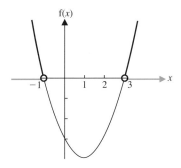

From the graph we see that the ranges of values of x which satisfy the given inequality are

$$x < -1 \quad \text{and} \quad x > 3$$

Note that the solution in the above example was two separate ranges each with its own boundary. If however we consider the inequality $(x - 3)(x + 1) < 0$ the part of the graph of $f(x)$ for which $f(x) < 0$ is below the x-axis and the corresponding values of x are $-1 < x < 3$. This time there is only one range, but it still has two boundaries.

EXERCISE 11c

Find the ranges of values of x that satisfy the following inequalities.

1. $(x-2)(x-1) > 0$

2. $(x+3)(x-5) \geqslant 0$

3. $(x-2)(x+4) < 0$

4. $(2x-1)(x+1) \geqslant 0$

5. $x^2 - 4x > 3$

6. $4x^2 < 1$

7. $(2-x)(x+4) \geqslant 0$

8. $5x^2 > 3x + 2$

9. $(3-2x)(x+5) \leqslant 0$

10. $(x-1)^2 > 9$

11. $(x+1)(x+2) \leqslant 4$

12. $(1-x)(4-x) > x + 11$

Problems

The types of problems which involve inequalities are very varied. Their solutions depend not only on all the methods used so far in this chapter but also on other facts known to the reader, for example:

1) a perfect square can never be negative.

2) the nature of the roots of a quadratic equation $ax^2 + bx + c = 0$ depends upon whether $b^2 - 4ac = 0$ or $b^2 - 4ac > 0$ or $b^2 - 4ac < 0$. As two of the above conditions are inequalities, many problems about the roots of a quadratic equation require the solution or interpretation of inequalities.

The worked examples that follow are intended to give the reader some ideas to use in problem solving and make no claim to cover every situation.

Examples 11d

1. Find the range(s) of values of k for which the roots of the equation $kx^2 + kx - 2 = 0$ are real.

$$kx^2 + kx - 2 = 0$$

For real roots '$b^2 - 4ac$' $\geqslant 0$

i.e. $\quad k^2 - 4(k)(-2) \geqslant 0$

$\Rightarrow \quad k(k+8) \geqslant 0$

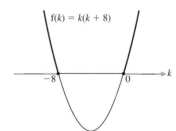
f(k) = k(k + 8)

From the sketch we see that
$k(k+8) \geqslant 0$ for $k \leqslant -8$ and $k \geqslant 0$

Therefore the equation $kx^2 + kx - 2 = 0$ has real roots if the value of k lies in either of the ranges $k \leqslant -8$ or $k \geqslant 0$

Note. This type of question is sometimes expressed in another, less obvious, way, i.e. 'If x is real and $kx^2 + kx - 2 = 0$, find the values that k can take'. Once the reader appreciates that, because x is real the roots of the equation are real, the solution is identical to that above.

2. Prove that $x^2 + 2xy + 2y^2$ cannot be negative.

Knowing that a perfect square cannot be negative, we rearrange the given expression in the form of perfect squares.

$$x^2 + 2xy + 2y^2 = x^2 + 2xy + y^2 + y^2$$
$$= (x+y)^2 + y^2$$

Each of the two terms on the RHS is a square and so cannot be negative.

Therefore $x^2 + 2xy + 2y^2$ cannot be negative.

EXERCISE 11d

1. Find the range(s) of values of k for which the given equation has real different roots

    (a) $kx^2 - 2x + k = 0$ (b) $x^2 + 3kx + k = 0$

2. Find the ranges of values of p for which the given equation has real roots.

    (a) $x^2 + (p+3)x + 4p = 0$ (b) $x^2 + 3x + 1 = px$

3. Find the range of values of a for which the equation $x^2 - ax + (a+3) = 0$ has no real roots.

4. Given that $px^2 + (p-3)x + 1 = 0$

    (a) find the range(s) of values of p for which the equation has real different roots

    (b) write down the values of p for which the equation has equal roots

    (c) write down the range(s) of values of p for which the roots of the equation are not real.

5. Show that $x^2 - 4xy + 5y^2 \geqslant 0$ for all real values of x and y

6. Prove that $(a+b)^2 \geqslant 4ab$ for all real values of a and b

MIXED EXERCISE 11

Solve each of the inequalities given in Questions 1 to 8.

1. $2x + 1 < 4 - x$ 2. $x - 5 > 1 - 3x$

3. $6x - 5 > 1 + 2x$ 4. $(x - 3)(x + 2) > 0$

5. $(2x - 3)(3x + 2) < 0$ 6. $x^2 - 3 < 10$

7. $(x - 3)^2 > 2$ 8. $(3 - x)(2 - x) < 20$

In each question from 9 to 12, find the set of values of x that satisfy the given inequalities.

9. $-3 < 5 - 2x < 3$ 10. $1 + 4x > 3x - 5 > 5x + 1$

11. $1 - 3x < 5 + x < 9 - x$ 12. $2x + 4 > 3x + 7 > 2x - 1$

13. Prove that $x^2 + y^2 - 10y + 25 \geqslant 0$ for all real values of x and y.

14. For what values of k does the equation $4x^2 + 8x - 8 = k(4x - 3)$ have real roots?

*15. (a) Draw a sketch of the graph of $y = x(x - 1)(x - 2)$. Hence find the range of values of x for which $x(x - 1)(x - 2) > 0$.

   (b) Use sketch graphs of $y = x^3$ and $y = x^2 - 1$ to find, approximately, the range of values of x for which $x^3 < x^2 - 1$. Use a graphics calculator or computer to find this range more accurately.

   (c) In order for the delivery of prepared fresh meals to take place on a given date, the raw ingredients have to be in the factory within d days before the given date. There are certain constraints on when the raw materials can be taken into the factory. Translating these constraints into mathematical terms gives the following inequality.

   $$(\tfrac{1}{2})(d - 3)^2 - 1 \leqslant \frac{1}{d}$$

   Within how many days before the given date for delivery of the finished product, do the raw ingredients have to be in the factory?

CHAPTER 12

MODULUS FUNCTIONS

The modulus of a number is its *positive* value, e.g. the modulus of -3 is 3 and the modulus of 3 is also 3. The modulus of any number x is written $|x|$.

(The modulus of x is also called the *absolute value* of x and it is written as abs x.)

THE MODULUS OF A FUNCTION

When $f(x) = x$, $f(x)$ is negative when x is negative.
But if $g(x) = |x|$, g takes the *positive* numerical value of x,
e.g. when $x = -3$, $|x| = 3$, so $g(x)$ is always positive.

Therefore the graph of $g(x) = |x|$ can be obtained from the graph of $f(x) = x$ by changing, to the equivalent positive values, the part of the graph of $f(x) = x$ for which $f(x)$ is negative.

Thus for negative values of $f(x)$, the graph of $g(x) = |x|$ is the reflection of $f(x) = x$ in the y-axis.

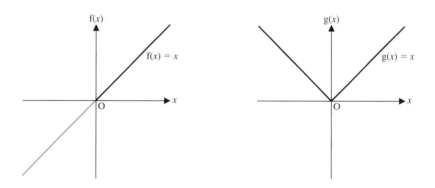

In general, the curve C_1 whose equation is $y = |f(x)|$ is obtained from the curve C_2 with equation $y = f(x)$, by reflecting in the x-axis the parts of C_2 for which $f(x)$ is negative. The remaining sections of C_1 are not changed.

For example, to sketch $y = |(x-1)(x-2)|$ we start by sketching the curve $y = (x-1)(x-2)$. We then reflect in the x-axis the part of this curve which is below the x-axis.

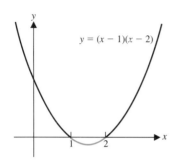 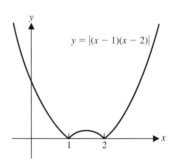

Note that for any function f, the mapping $x \rightarrow |f(x)|$ is also a function.

EXERCISE 12a

Sketch the following graphs. If you have one, use a graphics calculator to check your sketches.

1. $y = |x + 1|$ 2. $y = |x - 1|$

3. $y = |2x - 1|$ 4. $y = |x(x-1)(x-2)|$

5. $y = |x^2 - 1|$ 6. $y = |x^2 + 1|$

7. $y = |x^2 - x - 20|$ 8. $y = |(x+1)(x-2)|$

9. $y = \left| \dfrac{1}{x+1} \right|$ 10. $y = |x^3|$

The Effect of a Modulus Sign on a Cartesian Equation

When a section of the curve $y = f(x)$ is reflected in Oy, the equation of that part of the curve becomes $y = -f(x)$,

e.g., if $y = |x|$ for $x \in \mathbb{R}$
we can write this equation as

$$\begin{cases} y = x & \text{for } x \geqslant 0 \\ y = -x & \text{for } x < 0 \end{cases}$$

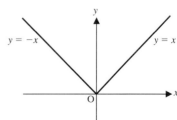

Intersection

To find the points of intersection between two graphs whose equations involve a modulus, we first sketch the graphs to locate the points roughly. Then we identify the equations in non-modulus form for each part of the graph. If these equations are written on the sketch then the correct pair of equations for solving simultaneously can be identified.

For example, the points common to $y = x - 1$ and $y = |x^2 - 3|$ can be seen from the sketch.

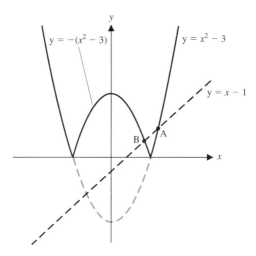

We can also see from the sketch that the coordinates of A satisfy the equations

$$y = x - 1 \quad \text{and} \quad y = x^2 - 3 \qquad [1]$$

and the coordinates of B satisfy the equations

$$y = x - 1 \quad \text{and} \quad y = 3 - x^2 \qquad [2]$$

Solving equations [1] gives $x^2 - x - 2 = 0 \quad \Rightarrow \quad x = -1$ or 2

It is clear from the diagram that $x \neq -1$, so A is the point $(2, 1)$

Similarly, solving equations [2] gives $x^2 + x - 4 = 0$

$$\Rightarrow \qquad x = -\tfrac{1}{2}\left(1 \pm \sqrt{17}\right)$$

Again from the diagram, it is clear that the x-coordinate of B is positive, so at B, $x = -\tfrac{1}{2}\left(1 - \sqrt{17}\right)$

Then using $y = x - 1$ gives $y = \tfrac{1}{2}\left(-3 + \sqrt{17}\right)$

This example illustrates the importance of checking solutions to see if they are relevant to the given problem.

Solution of Equations Involving a Modulus Sign

An equation such as $1 - |x| = 3x - 1$ can be solved by finding the values of x at the points of intersection of two graphs. As we are dealing with an equation, we can rearrange it if necessary to give graphs that are simple as possible.

Example 12b

Solve the equation $1 - |x| = 3x - 1$

$$1 - |x| = 3x - 1 \quad \Rightarrow \quad |x| = 2 - 3x$$

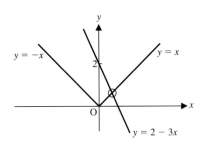

A sketch of the graphs $y = |x|$ and $y = 2 - 3x$ shows that $|x| = 2 - 3x$ where $x = 2 - 3x$

$$x = 2 - 3x \quad \Rightarrow \quad x = \tfrac{1}{2}$$
$$\therefore \qquad 1 - |x| = 3x - 1 \quad \text{when} \quad x = \tfrac{1}{2}$$

EXERCISE 12b

Find the points of intersection of the graphs.

1. $y = |x|$ and $y = x^2$

2. $y = x$ and $y = |x^2 - 1|$

3. $y = |1/x|$ and $y = |x|$

4. $y = 9 - x^2$ and $y = |3x - 1|$

5. $y = |x^2 - 4|$ and $y = 2x + 1$

6. $y = |x^2 - 1|$ and $y = |x - 2|$

Solve the equations

7. $2|x| = 2x + 1$

8. $|x + 1| = |x - 1|$

9. $|2x - 1| + 2 = 4x$

10. $|x^2 - 1| - 1 = 3x - 2$

11. $|2 - x^2| + 2x + 1 = 0$

12. $4 - |x^2 - 4| = 5x$

Graphs of Functions Involving a Modulus

Graphs of functions involving one modulus sign can often be sketched easily by building up the 'picture' using transformations. It helps to start with a sketch of the graph inside the modulus function.

Examples 12c

1. Sketch the graph of $y = 3 - |1 - 2x|$

    We can use transformations to build up the picture in stages.

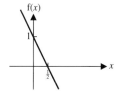

    1)   Draw $f(x) = 1 - 2x$

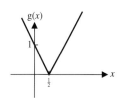

    2)   Draw $g(x) = |1 - 2x|$

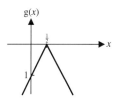

    3)   Draw $-|1 - 2x|$
            $-g(x)$ is the reflection of $g(x)$ in Ox.

    4) $y = 3 - |1 - 2x|$ is the graph of $-g(x)$ translated 3 units upwards.

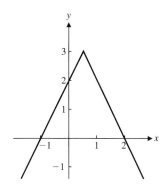

Inequalities involving modulus functions can be solved easily from sketch graphs. Sometimes a minor rearrangement of the inequality helps; aim to have not more than one modulus function on each side.

2. **Find the set of values of** x **for which** $|x-3|+|x|<5$

$$|x-3|+|x|<5 \quad \Rightarrow \quad |x-3|<5-|x|$$

We can now sketch the graphs $y=|x-3|$ and $y=5-|x|$.

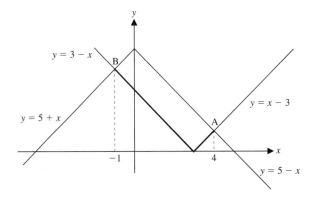

At A $\qquad x-3=5-x \quad \Rightarrow \quad x=4$

and at B $\qquad 3-x=5+x \quad \Rightarrow \quad x=-1$

From the graph we see that $|3-x|<5-|x|$ for $-1<x<4$

EXERCISE 12c

In questions 1 to 4, sketch the graphs of the functions.

1. $y=|x+2|-3$

2. $y=2-|x-1|$

3. $y=|x^2-1|-1$

4. $y=1-|x^2-2|$

Solve the following inequalities.

5 $|x| < 1 - |x|$

9. $3x - 1 < 1 + |x|$

6. $|x - 1| < |x + 2|$

10. $|3x + 2| + |x + 1| > 2$

7. $2|x| < |1 - x|$

11. $|1 - x^2| < 2x + 1$

8. $|x + 1| < 2x$

12. $|x| - 1 > 1 - x^2$

13. The diagram shows the graph of $y = |f(x)|$

Draw separate sketches showing possible graphs of $y = f(x)$.

(a)   (b)

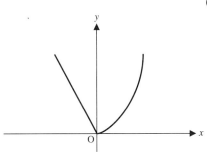 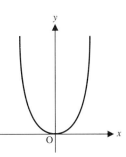

14.

Two yachts sail towards and then past each other in a direct line. When they start they are 50 nautical miles apart. One yacht sails at an average speed of 5 knots (nautical miles per hour) and the other yacht sails at an average speed of 8 knots. The two yachts can maintain radio contact with each other provided they are less than 20 nautical miles apart. Assuming that their average speeds are constant speeds, find an expression for the distance, d nautical miles, between the yachts in terms of the time, t hours, for which they have been sailing. Hence find for how long they can maintain radio contact.

CONSOLIDATION C

SUMMARY

CIRCLE THEOREMS

The perpendicular bisector of any chord goes through O, and conversely.

$\angle O = 2\angle P$

$\angle P = \angle Q$

$\angle P = 90°$

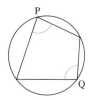

$\angle P + \angle Q = 180°$

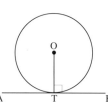

The tangent at T is perpendicular to OT

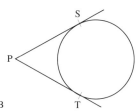

$PS = PT$

CIRCULAR MEASURE

One radian (1^c) is the size of the angle subtended at the centre of a circle by an arc equal in length to the radius of the circle.

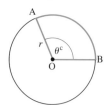

The length of arc AB is $r\theta$

The area of sector AOB is $\frac{1}{2}r^2\theta$

INEQUALITIES

If $a > b$ then $a + k > b + k$ for all values of k

$$ak > bk \quad \text{for all positive values of } k$$

$$ak < bk \quad \text{for all negative values of } k$$

FUNCTIONS

A function f is a rule that maps a number x to another single number $f(x)$. The domain of a function is the set of input numbers, i.e. the set of values of x.

The range of a function is the set of output values, i.e. the set of values of $f(x)$.

The general form of a quadratic function is $f(x) = ax^2 + bx + c$ where $a \neq 0$

If $a > 0$, $f(x)$ has a minimum value where $x = -b/2a$
If $a < 0$, $f(x)$ has a maximum value where $x = -b/2a$

A function is *even* if $f(x) = f(-x)$
Even functions are symmetrical about the y-axis

A function is *odd* if $f(x) = -f(-x)$
Odd functions have rotational symmetry about the origin.

The function that maps the output of f to its input is called the inverse function of f, and is denoted by f^{-1}, i.e. $f^{-1} : f(x) \to x$

Note that while it is always possible to reverse a mapping, the rule that does this may not be a function, so not all functions have an inverse.

When a function f operates on a function g we have a function of a function, or a composite function, which is denoted by fg, or $f \circ g$

TRANSFORMATIONS OF CURVES

$y = f(x) + c$ is a translation of $y = f(x)$ by c units in the direction Oy

$y = f(x + c)$ is a translation of $y = f(x)$ by c units in the direction xO

$y = -f(x)$ is the reflection of $y = f(x)$ in the x-axis

$y = f(-x)$ is the reflection of $y = f(x)$ in the y-axis

Modulus Functions

The modulus of x is written $|x|$ and it means the positive numerical value of x, e.g. when $x = -3$, $|x| = 3$

The modulus of x is also called the absolute value of x, abs x.

The graph of $y = |f(x)|$ is obtained from the graph of $y = f(x)$ by reflecting in the x-axis the parts of the graph for which $f(x)$ is negative. The sections for which $f(x)$ is positive remain unchanged.

MULTIPLE CHOICE EXERCISE C

TYPE I

1. The least value of $(x-1)(x-3)$ is when x is equal to

   A   1   **B**   3   **C**   2   **D**   0   **E** -2

2. If $f(x) = 2x - 1$ then $f^{-1}(x)$ is

   A $1 - 2x$ **C** $2y - 1$ **E** $2x + 1$

   B $\frac{1}{2}(x+1)$ **D** $\frac{1}{2}x - 1$

3. The values of x for which $(x-1)(x-5) < 0$ are

   A $x < 3$ **C** $x < 0$ **E** $1 \leqslant x \leqslant 5$

   B $x < 1, \ x > 5$ **D** $1 < x < 5$

4. An angle of 1 radian is equivalent to:

   A $90°$ **B** $60°$ **C** $\approx 67.3°$ **D** $\approx 57.3°$ **E** $45°$

5. An arc PQ subtends an angle of $60°$ at the centre of a circle of radius 1 cm. The length of the arc PQ is

   A   60 cm   **B**   30 cm   **C** $\frac{1}{6}\pi$ cm   **D** $\frac{1}{3}\pi$ cm   **E** $\frac{1}{18}\pi^2$ cm

6. The graph of $f(x) = 1 - |x|$ could be

   A   **C**   **E**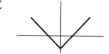

   B   **D**

7. If $|x| = |2 - x|$ then x is

   A   0   **B**   2   **C**   1 and -1 **D** -1 **E**   1

8. The curve $y = f(x)$ is

A

C

E

B

D

The curve $y = f(-x)$ could be

9. The equation of a curve C is $y = ax^2 + 1$, where a is a constant. Which of these curves cannot be C?

A

C

E

B

D

10. The graph of $y = |x + 1|$ could be

A
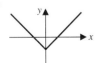

B

C

D

11. The graph of $y = f(x)$ is

The graph of $y = f(x + a)$ could be

A   B   C   D

  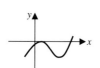

TYPE II

12. π can represent half the circumference of a circle.

13. $f(x) = x^2$, $x \in \mathbb{R}$, f is an even function.

14. $x^2 - 4 < 0$ when $x < 4$

15. If $f(x) = |x - 2|$, $f(1) = 3$

16. If an arc of a circle of radius 0.5 cm subtends an angle of $60°$ at the centre of the circle, then the length of the arc is 30 cm.

17. If $\dfrac{1}{x} < 2$ then $\dfrac{1}{2} < x$

18. If $x - a$ is a factor of $x^2 + px + q$, the equation $x^2 + px + q = 0$ has a root equal to a.

19. If $f : x \to 2x - 1$ and $g : x \to x^2$ then $fg : x \to (2x - 1)^2$

20. $|x + 1| \equiv |x| - 1$

21. In the interval $0 < x < 1$, $|x + 1| > 0$

MISCELLANEOUS EXERCISE C

1.

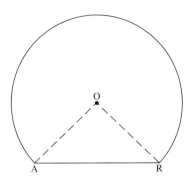

The diagram shows the cross-section of a tunnel. The cross-section has the shape of a major segment of a circle, and the point O is the centre of the circle. The radius of the circle is $4\,\mathrm{m}$, and the size of angle AOB is 1.5 radians. Calculate the perimeter of the cross-section.   (UCLES)$_s$

2. Find the set of values of x for which

$$2(x^2 - 5) < x^2 + 6.$$ (NEAB)

3. Find the set of values of x for which

$$x^2 - x - 12 > 0.$$ (ULEAC)

4.

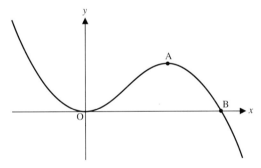

The diagram shows the graph of $y = x^2(3 - x)$. The coordinates of the points A and B on the graph are $(2, 4)$ and $(3, 0)$ respectively.

(a) Write down the solution set of the inequality $x^2(3 - x) \leqslant 0$.

(b) The equation $3x^2 - x^3 = k$ has three real solutions for x. Write down the set of possible values for k.

(c) Functions f and g are defined as follows:

$$\mathrm{f} : x \mapsto x^2(3 - x), \ 0 \leqslant x \leqslant 2,$$
$$\mathrm{g} : x \mapsto x^2(3 - x), \ 0 \leqslant x \leqslant 3$$

Explain why f has an inverse while g does not.

(d) State the domain and range of f^{-1}, and sketch the graph of f^{-1}.   (UCLES)$_s$

5. A piece of wire of length 4 metre is bent into the shape of a sector of a circle of radius r metre and angle θ radian.

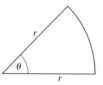

(a) State, in terms of θ and r.
(i) the length of the arc,
(ii) the area A of the sector.
(b) Hence show that $A = 2r - r^2$.
(c) Find the value of r which will make the area a maximum. Deduce the corresponding value of θ.
(d) The figures labelled A–F below show, all to the same scale, six possible sectors which can be made from the piece of wire. Which of them has the largest area?

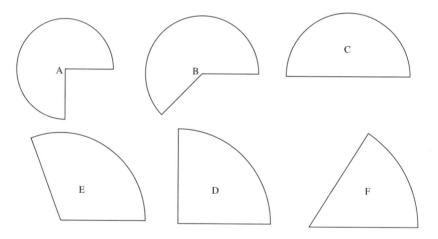

6. The function f with domain $\{x : x \leqslant 0\}$ is defined by $f(x) = \dfrac{8}{x+2}$.

(a) Sketch the graph of f and state the range of f.
(b) Find $f^{-1}(x)$, where f^{-1} denotes the inverse of f.
(c) Calculate the value of x for which $f(x) = f^{-1}(x)$. (AEB)

7. A landscape gardener is given the following instructions about laying a rectangular lawn. The length x m is to be 2 m longer than the width. The width must be greater than 6.4 m and the area is to be less than 63 m². By forming an inequality in x, find the set of possible values of x. (ULEAC)ₛ

8. Express $2x^2 + 5x + 4$ in the form $a(x+b)^2 + c$ stating the numerical values of a, b and c.

Hence, or otherwise, write down the coordinates of the minimum point on the graph of $y = 2x^2 + 5x + 4$. (UCLES)

9. The function f is given by

$$f : x \mapsto x^2 - 8x, \quad x \in \mathbb{R}, \quad x \leqslant 4.$$

(a) Determine the range of f. (b) Find the value of x for which $f(x) = 20$.

(c) Find $f^{-1}(x)$ in terms of x. (ULEAC)

10. The function f is defined by

$$f : x \rightarrow \frac{3x+1}{x-2}, \quad x \in \mathbb{R}, \quad x \neq 2$$

Find, in a similar form, the functions

(a) ff (b) f^{-1} (ULEAC)

11.

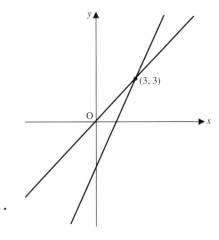

The diagram shows the graphs of the lines $y = x$ and $y = 2x - 3$, and their point of intersection $(3, 3)$. Sketch on a single diagram the graphs of $y = |x|$ and $y = |2x - 3|$, and give the points of intersection of these graphs. (UCLES)ₛ

12. The function f has domain the set of all non-zero real numbers, and is given by $f(x) = \frac{1}{x}$ for all x in this set. On a single diagram, sketch each of the following graphs, and indicate the geometrical relationships between them.

(a) $y = f(x)$, (b) $y = f(x+1)$, (c) $y = f(x+1) + 2$.

Deduce, explaining your reasoning, the coordinates of the point about which the graph of $y = \frac{2x+3}{x+1}$ is symmetrical. (UCLES)ₛ

13.

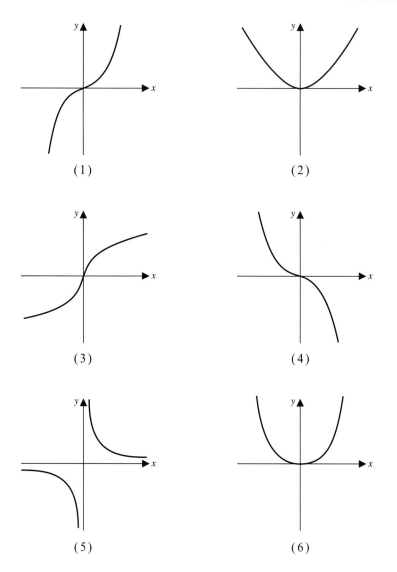

(1)   (2)

(3)   (4)

(5)   (6)

The figures (1), (2), ..., (6) show the graphs of six functions.

(a)   Two of the figures show the graphs of a function and its inverse function. State which these two figures are.

(b)   Two of the figures show the graphs of even function. State which these two figures are.

(c)   Two of the figures are such that, while one shows the graph of $y = g(x)$, the other shows the graph of $y = -g(x)$. State which these two figures are.

(OCSEB)$_{(p)}$

14.

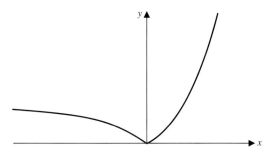

The diagram shows the graph of $y = |f(x)|$, for a certain function f with domain $\mathbb{R}$. Sketch, on separate diagrams, two possibilities for the graph of $y = f(x)$.   (UCLES)$_s$

15. The function f and g are defined by

$$f : x \mapsto 3x - 3, \quad x \in \mathbb{R},$$
$$g : x \mapsto x^2 + 1, \quad x \in \mathbb{R},$$

(a) Find the range of g.

(b) Calculate the value of $gf(2)$.

(c) Determine the value of x for which $gf(x) = fg(x)$.

(d) Determine the values of x for which $|f(x)| = 8$.   (ULEAC)$_s$

16. The functions f and g are defined by

$$f : x \mapsto x^2 - 10, \quad x \in \mathbb{R},$$
$$g : x \mapsto |x - 2|, \quad x \in \mathbb{R},$$

(a) Show that $f \circ f : x \mapsto x^4 - 20x^2 + 90, \quad x \in \mathbb{R}$.
    Find all the values of x for which $f \circ f(x) = 26$.

(b) Show that $g \circ f(x) = |x^2 - 12|$. Sketch a graph of $g \circ f$.
    Hence, or otherwise, solve the equation $g \circ f(x) = x$.   (AEB)$_s$

17. Solve the inequality $|x + 1| < |x - 2|$.   (UCLES)$_s$

18. Sketch the graph of $y = |x + 3|$ and hence solve the inequality $|x + 3| < x^2$.

19. An arc of a circle, radius r cm, subtends an angle of exactly 1.2 radians at the centre of the circle. The length s cm of the arc is calculated using $r = 9$. Given that this value of r has been rounded to the nearest integer, determine the maximum absolute error in the calculated value of s.   (ULEAC)$_{ps}$

20. (a) Write down an example of a polynomial in x of order 4.

(b) In an experiment, Ama measures the value of s at different times, t. Her results are shown as the curve on the graph below

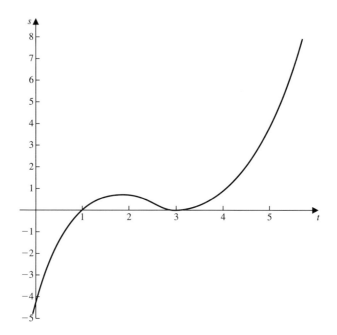

Ama believes that it is possible to model s as a polynomial in t.

(i) Explain why it is reasonable to think that the order of such a polynomial might be 3.
Ama proposes a model of the form

$$s = a(t-p)(t-q)^2$$

(ii) Write down the points where the curve meets the coordinate axes and use them to find values for p, q and a.

(iii) Compare the values obtained from the model with those on the graph when $t = 2$, 4 and 5, and comment on the quality of the model.
Ama proposes a refinement to the model making it into

$$s = a(t-p)(t-q)^2(1-ht)$$

where a, p and q have the same values as before and h is a small positive constant. Ama chooses the value of h so that the model and the graph are in agreement when $t = 5$.

(iv) Find the value of h.   (MEI)$_s$

21. The diagram shows the graph of the curve $y = f(x)$.

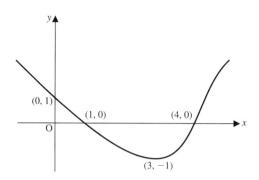

Sketch, showing the coordinates of the intercepts on the axes and the turning point, the curve whose equation is

(a) $y = |f(x)|$ (b) $y = f(x) + 1$ (c) $y = f(x+4)$ (d) $y = 2f(x)$

22. The shape of the earring pendant, shown shaded in the diagram, is obtained from two equal overlapping circles such that height of the pendant, which is 2 cm, is equal to the radius of the circles. Find the area of the pendant.

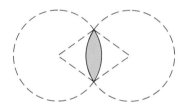

CHAPTER 13

DIFFERENTIATION

CHORDS AND TANGENTS

Consider any two points, A and B, on any curve.

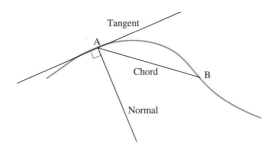

The line joining A and B is called a chord.

The line that touches the curve at A is called the tangent at A.

Note that the word *touch* has a precise mathematical meaning, i.e. a line that meets a curve at a point and carries on without crossing to the other side of the curve at that point, is said to *touch* the curve at the *point of contact*.

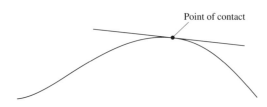

The line perpendicular to the tangent at A is called the normal at A.

THE GRADIENT OF A CURVE

Gradient, or slope, defines the direction of a line (lines can be straight or curved).

When walking along a straight line we walk in the same direction all the time, i.e. the gradient of a straight line is constant.

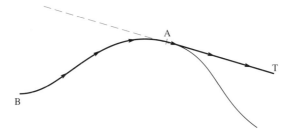

If, however, we move from B to A along a curve, our direction is changing all the time, i.e. the gradient of a curve is not constant but has different values at different points on the curve.

Now if at A we continue to move, but without any further change in direction, we go along the straight line AT which is the tangent at A, so

**the gradient of a curve at a point A is the same as
the gradient of the tangent at A**

For a straight line the numerical value of the gradient is found by taking the coordinates of any two points on the line and working out

$$\frac{\text{increase in } y}{\text{increase in } x}$$

This can be used to find the gradient of a tangent to a curve but, if the tangent is just drawn by eye, the value obtained can only be an approximation.

A precise method is needed for finding the gradient of a curve whose equation is known, so that further analysis can be made of the properties of the curve.

Let us consider the problem of finding the gradient of the tangent at a point A on a curve.

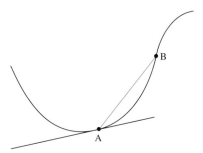

If B is another point on the curve, fairly close to A, then the gradient of the chord AB gives an *approximate* value for the gradient of the tangent at A. As B gets nearer to A, the chord AB gets closer to the tangent at A, so the approximation becomes more accurate.

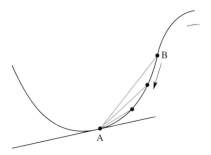

So, as B gets closer and closer to A, we can say,

the gradient of chord AB → the gradient of the tangent at A

This fact can also be expressed in the form

$$\underset{\text{as } B \to A}{\textbf{limit}}\ (\textbf{gradient of chord AB}) = \textbf{gradient of tangent at A}$$

This definition can be applied to a particular point on a particular curve. Suppose, for instance, that we want the gradient of the curve $y = x^2$ at the point where $x = 1$

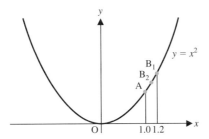

A is the point $(1, 1)$ and we need a succession of points, B_1, B_2, ... getting closer and closer to A. Let us take the points where $x = 1.2$, 1.1, 1.05, 1.01, 1.001, then calculate the corresponding y coordinates, find the increases in x and y between A and B and hence find the gradient of AB.

x	1.2	1.1	1.05	1.01	1.001
y $(= x^2)$	1.44	1.21	1.1025	1.0201	1.002 001
Increase in y from A to B	0.44	0.21	0.1025	0.0201	0.002 001
Increase in x from A to B	0.2	0.1	0.05	0.01	0.001
Gradient of chord AB	2.2	2.1	2.05	2.01	2.001

From the numbers in the last row of the table it is clear that, as B gets nearer to A, the gradient of the chord gets nearer to 2, i.e.

$$\underset{\text{as } B \to A}{\text{limit}} \ (\text{gradient of chord AB}) = 2$$

It is equally clear that it is much too tedious to go through this process each time we want the gradient at just one point on just one curve and that we need a more general method. For this we use a general point $A(x, y)$ and a variable small change in the value of x between A and B.

A new symbol, δ, is used to denote this small change.

When δ appears as a prefix to any letter representing a variable quantity, it denotes a small increase in that quantity,

e.g. δx means a small increase in x
δy means a small increase in y
δt means a small increase in t

Note that δ is only a prefix. It does not have an independent value and cannot be treated as a factor.

Now consider again the gradient of the curve with equation $y = x^2$

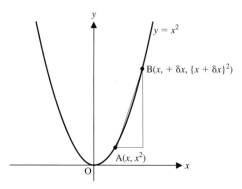

This time we will look for the gradient at *any* point $A(x, y)$ on the curve and use a point B where the x-coordinate of B is $x + \delta x$.

For any point on the curve, $y = x^2$

So, at B, the y-coordinate is $(x + \delta x)^2 = x^2 + 2x\delta x + (\delta x)^2$

Therefore the gradient of chord AB, which is given by $\dfrac{\text{increase in } y}{\text{increase in } x}$, is

$$\frac{(x + \delta x)^2 - x^2}{(x + \delta x) - x} = \frac{2x\delta x + (\delta x)^2}{\delta x} = 2x + \delta x$$

Now, as $B \to A$, $\delta x \to 0$, therefore

$$\text{gradient of curve at A} = \underset{\text{as } B \to A}{\text{limit}} \ (\text{gradient of chord AB})$$

$$= \underset{\text{as } \delta x \to 0}{\text{limit}} \ (\text{gradient of chord AB})$$

$$= \underset{\text{as } \delta x \to 0}{\text{limit}} \ (2x + \delta x)$$

$$= 2x$$

This result can now be used to give the gradient at any point on the curve with equation $y = x^2$, where the x-coordinate is given, e.g.

at the point where $x = 3$, the gradient is $2(3) = 6$
and at the point $(4, 16)$, the gradient is $2(4) = 8$

Looking back at the longer method we used on page 197 to find the gradient at the point where $x = 1$, we see that the value obtained there is confirmed by using the general result,
i.e. gradient $= 2x = 2(1) = 2$

DIFFERENTIATION

The process of finding a general expression for the gradient of a curve at any point is known as differentiation.

The general gradient expression for a curve $y = f(x)$ is itself a function so it is called the *gradient function*. For the curve $y = x^2$ for example, the gradient function is $2x$.

Because the gradient function is derived from the given function, it is more often called the *derived function* or the *derivative*.

The method used above, in which the limit of the gradient of a chord was used to find the derived function, is known as *differentiating from first principles*. It is the fundamental way in which the gradient of each new type of function is found and, although many short cuts can be developed, it is important to understand this basic method, which can be applied to any function of x.

The General Gradient Function

Consider any curve with equation $y = f(x)$.

Taking two points on the curve, $A(x, y)$ and $B(x + \delta x, y + \delta y)$ we have,

at A, $y = f(x)$ and at B, $y + \delta y = f(x + \delta x)$

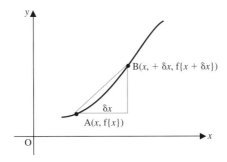

The gradient of AB is given by

$$\frac{\delta y}{\delta x} = \frac{f(x + \delta x) - f(x)}{\delta x}$$

Therefore **the gradient of the curve at A is given by**

$$\underset{\delta x \to 0}{\text{limit}} \frac{\delta y}{\delta x} = \underset{\delta x \to 0}{\text{limit}} \frac{f(x + \delta x) - f(x)}{\delta x}$$

Notation

The gradient function of a curve $y = f(x)$ can be denoted by $\dfrac{dy}{dx}$
(we say dy by dx)

i.e. $$\frac{dy}{dx} = \lim_{\delta x \to 0} \frac{\delta y}{\delta x} = \lim_{\delta x \to 0} \frac{f(x + \delta x) - f(x)}{\delta x}$$

Note that d has no independent meaning and must never be regarded as a factor.

The complete symbol $\dfrac{d}{dx}$ means 'the derivative with respect to x of'

So $\dfrac{dy}{dx}$ means 'the derivative with respect to x of y'

(Because the phrase 'with respect to' is used frequently, it is often abbreviated to w.r.t.)

We have seen that the derivative of x^2 is $2x$ so we can write

for $y = x^2$, $\dfrac{dy}{dx} = 2x$

An alternative notation concentrates on the function of x rather than the equation of the curve. Using this form we can write

for $f(x) = x^2$, $f'(x) = 2x$

In this form, f' means 'the gradient function' or 'the derived function'.

DIFFERENTIATING x^n WITH RESPECT TO x

We have shown that $\dfrac{d}{dx}(x^2) = 2x$. By differentiating from first principles it

can also be shown that $\dfrac{d}{dx}(x^3) = 3x^2$ and that $\dfrac{d}{dx}(x^{-1}) = -x^{-2}$

The pattern appears to be that when we differentiate a power of x we multiply by the power and then reduce the power by 1, i.e. it looks as though

if $y = x^n$, then $\dfrac{dy}{dx} = nx^{n-1}$

This result, although deduced from just a few examples, is in fact valid for all powers, including those that are fractional or negative, but it is not possible to give a proof at this stage. It is easy to apply and makes the task of differentiating a power of x very much simpler, e.g.

$$\frac{d}{dx}(x^7) = 7x^6, \qquad \frac{d}{dx}(x^{-2}) = -2x^{-3}, \qquad \frac{d}{dx}(x^{3/2}) = (3/2)x^{1/2}$$

Example 13a

Differentiate with respect to x (a) $\sqrt{x^3}$ (b) $x^{-1/3}$

(a)   To use the rule we must express $\sqrt{x^3}$ in the form x^n, i.e. $\sqrt{x^3} = x^{3/2}$

Using $\dfrac{d}{dx}(x^n) = nx^{n-1}$ where $n = \frac{3}{2}$ gives

$$\frac{d}{dx}(x^{3/2}) = \frac{3}{2}x^{(3/2-1)} = \frac{3}{2}x^{1/2} \text{ or } \frac{3}{2}\sqrt{x}$$

(b)   This time $n = -\frac{1}{3}$ so

$$\frac{d}{dx}(x^{-1/3}) = -\frac{1}{3}x^{(-1/3-1)} = -\frac{1}{3}x^{-4/3}$$

EXERCISE 13a

Differentiate with respect to x

1. x^5 2. x^{11} 3. x^{20} 4. x^{10} 5. x^{-3}

6. x^{-7} 7. x^{-5} 8. $x^{4/3}$ 9. x 10. x^{-1}

11. $x^{1/3}$ 12. $\sqrt{x}$ 13. $\dfrac{1}{x^2}$ 14. $\sqrt{\dfrac{1}{x}}$ 15. $\dfrac{1}{x^4}$

16. $\dfrac{1}{x^{10}}$ 17. $x^{-1/4}$ 18. $\sqrt{x^5}$ 19. $x^{1/7}$ 20. x^p

Differentiating a Constant

Consider the equation $y = c$.
Graphically this is a horizontal straight line and its gradient is zero,

i.e. $\dfrac{dy}{dx} = 0$

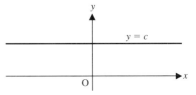

$$\text{If } y = c \quad \text{then} \quad \frac{dy}{dx} = 0$$

Differentiating a Linear Function of x

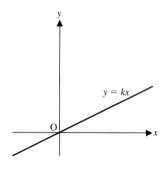

The graph of the equation $y = kx$,
where k is a constant,
is a straight line with gradient k.

Hence $\dfrac{d}{dx}(kx) = k$

Now if we apply the general rule for differentiating x^n to $y = x$,

i.e. to $y = x^1$, we get $\dfrac{d}{dx}(x^1) = 1x^0 = 1$

Combining these two facts shows that

$$\frac{d}{dx}(kx) = k \times \frac{d}{dx}(x)$$

This conclusion applies, in fact, to a constant multiple of *any* function of x,

e.g.   if $y = 3x^5$, $\dfrac{dy}{dx} = 3 \times \dfrac{d}{dx}(x^5) = 3 \times 5x^4 = 15x^4$

and if $y = 4x^{-2}$, $\dfrac{dy}{dx} = 4 \times -2x^{-3} = -8x^{-3}$

In general then, if a is a constant,

$$\frac{d}{dx}(ax^n) = anx^{n-1}$$

This rule, although not proved for the general case, can be used freely.

Another very useful property is that a function of x which contains the sum or difference of a number of terms can be differentiated term by term, applying the basic rule to each in turn, i.e.

$$\frac{d}{dx}[f(x)+g(x)] = \frac{d}{dx}f(x)+\frac{d}{dx}g(x)$$

Example 13b

If $y = x^4 + \dfrac{1}{x} - 6x$, find $\dfrac{dy}{dx}$

$$x^4 - \frac{1}{x} - 6x = x^4 - x^{-1} - 6x$$

$$\frac{dy}{dx} = \frac{d}{dx}(x^4) + \frac{d}{dx}(x^{-1}) - \frac{d}{dx}(6x) = 4x^3 - \frac{1}{x^2} - 6$$

EXERCISE 13b

Differentiate each of the following functions w.r.t. x

1. $5x^3$

2. $7x$

3. $\dfrac{8}{x}$

4. $5\sqrt{x}$

5. $\dfrac{-1}{2x^3}$

6. $3x^3 - 4x^2$

7. $4x - \dfrac{4}{x}$

8. $\dfrac{3}{x} + \dfrac{x}{3}$

9. $x^3 - x^2 + 5x - 6$

10. $3x^2 + 7 - 4/x$

11. $x^3 - 2x^2 - 8x$

12. $2x^4 - 4x^2$

13. $x^2 + 5\sqrt{x}$

14. $3x^3 - 4x^2 + 9x - 10$

15. $x^{3/2} - x^{1/2} + x^{-1/2}$

16. $\sqrt{x} + \sqrt{x^3}$

17. $\dfrac{1}{x^2} - \dfrac{1}{x^3}$

18. $\dfrac{1}{\sqrt{x}} - \dfrac{2}{x}$

19. $x^{-1/2} + 3x^{3/2}$

20. $x^{1/4} - x^{1/5}$

21. $\dfrac{4}{x^3} + \dfrac{x^3}{4}$

22. $\dfrac{4}{x} + \dfrac{5}{x^2} - \dfrac{6}{x^3}$

23. $3\sqrt{x} - 3x$

24. $x - 2x^{-1} - 3x^{-3}$

Differentiating Products and Fractions

All the rules given above can be applied to the differentiation of expressions containing products or quotients provided that they are multiplied out or divided into separate terms.

Examples 13c

1. If $y = (x-3)(x^2+7x-1)$, find $\dfrac{dy}{dx}$

$$y = (x-3)(x^2+7x-1) = x^3 + 4x^2 - 22x + 3$$

$$\Rightarrow \qquad \frac{dy}{dx} = 3x^2 + 8x - 22$$

2. Find $\dfrac{dt}{dz}$ given that $t = \dfrac{6z^2+z-4}{2z}$

$$t = \frac{6z^2+z-4}{2z} = \frac{6z^2}{2z} + \frac{z}{2z} - \frac{4}{2z}$$

$$= 3z + \frac{1}{2} - \frac{2}{z}$$

$$\Rightarrow \qquad \frac{dt}{dz} = 3 + 0 - 2(-z^{-2})$$

$$= 3 + \frac{2}{z^2}$$

EXERCISE 13c

In each question, differentiate with respect to the variable concerned.

1. $y = (x+1)^2$

2. $z = x^{-2}(2-x)$

3. $y = (3x-4)(x+5)$

4. $y = (4-z)^2$

5. $s = \dfrac{t^{-1}+3t^2}{2t^2}$

6. $s = \dfrac{t^2+t}{2t}$

7. $y = \left(\dfrac{1}{x}\right)(x^2+1)$

8. $y = \dfrac{z^3-z}{\sqrt{z}}$

9. $y = 2x(3x^2-4)$

10. $s = (t+2)(t-2)$

11. $s = \dfrac{t^3-2t^2+7t}{t^2}$

12. $y = \dfrac{\sqrt{x}+7}{x^2}$

THE GRADIENT OF A TANGENT

If the equation of a curve is known, and the gradient function can be found, then the gradient, m say, at a particular point A on that curve can be calculated. This is also the gradient of the tangent to the curve at A.

Examples 13d

1. The equation of a curve is $s = 6 - 3t - 4t^2 - t^3$. Find the gradient of the tangent to the curve at the point $(-2, 4)$.

$$s = 6 - 3t - 4t^2 - t^3 \quad \Rightarrow \quad \frac{ds}{dt} = 0 - 3 - 8t - 3t^2$$

At the point $(-2, 4)$, $\quad \dfrac{ds}{dt} = -3 - 8(-2) - 3(-2)^2 = 1$

Therefore the gradient of the tangent at $(-2, 4)$ is 1.

2. Find the coordinates of the points on the curve $y = 2x^3 - 3x^2 - 8x + 7$ where the gradient is 4.

$$y = 2x^3 - 3x^2 - 8x + 7 \quad \Rightarrow \quad \frac{dy}{dx} = 6x^2 - 6x - 8$$

If the gradient is 4 then $\dfrac{dy}{dx} = 4$

i.e. $6x^2 - 6x - 8 = 4 \quad \Rightarrow \quad 6x^2 - 6x - 12 = 0$

$\Rightarrow \quad x^2 - x - 2 = 0$

$\therefore \quad (x - 2)(x + 1) = 0 \quad \Rightarrow \quad x = 2 \text{ or } -1$

When $x = 2$, $y = 16 - 12 - 16 + 7 = -5$

when $x = -1$, $y = -2 - 3 + 8 + 7 = 10$

Therefore the gradient is 4 at the points $(2, -5)$ and $(-1, 10)$

EXERCISE 13d

Find the gradient of the tangent at the given point on the given curve.

1. $y = x^2 + 4$ where $x = 1$

2. $y = 3/x$ where $x = -3$

3. $y = \sqrt{z}$ where $z = 4$

4. $s = 2t^3$ where $t = -1$

5. $v = 2 - 1/u$ where $u = 1$

6. $y = (x+3)(x-4)$ where $x = 3$

7. $y = z^3 - z$ where $z = 2$

8. $s = t + 3t^2$ where $t = -2$

9. $z = x^2 - 2/x$ where $x = 1$

10. $y = \sqrt{x} + 1/\sqrt{x}$ where $x = 9$

11. $s = \sqrt{t}(1 + \sqrt{t})$ where $t = 4$

12. $y = \dfrac{x^2 - 4}{x}$ where $x = -2$

Find the coordinates of the point (s) on the given curve where the gradient has the value specified.

13. $y = 3 - \dfrac{2}{x};\ \dfrac{1}{2}$

14. $z = x^2 - x^3;\ -1$

15. $s = t^3 - 12t + 9;\ 15$

16. $v = u + \dfrac{1}{u};\ 0$

17. $s = (t+3)(t-5);\ 0$

18. $y = \dfrac{1}{x^2};\ \dfrac{1}{4}$

19. $y = (2x-5)(x+1);\ -3$

20. $y = z^3 - 3z;\ 0$

MIXED EXERCISE 13

1. Find $f'(x)$ when $f(x)$ is

(a) x^3 (b) $2x^2$ (c) $\dfrac{1}{2x}$ (d) $\dfrac{2}{\sqrt{x}}$ (e) $\dfrac{1}{4x^2}$ (f) $4\sqrt{x}$

2. Differentiate

(a) $3x^2 + x$ (b) $x^3 - 3x + 1$ (c) $(x-1)(x+2)$

3. Find the derivative of

(a) $x^{-3} - x^3 + 7$ (b) $x^{1/2} - x^{-1/2}$ (c) $1/x^2 + 2/x^3$

4. Differentiate w.r.t. x.

(a) $y = x^{3/2} - x^{2/3} + x^{-1/3}$ (b) $y = \sqrt{x} - 1/x + 1/x^3$

(c) $\dfrac{\sqrt{x}-1}{x}$ (d) $\dfrac{x-x^2}{\sqrt{x}}$

5. Find the gradient of the curve $y = 2x^3 - 3x^2 + 5x - 1$ at the point

(a) $(0, -1)$ (b) $(1, 3)$ (c) $(-1, -11)$

6. Find the gradient of the given curve at the given point.

(a) $y = x^2 + x - 9; \ x = 2$
(b) $y = x(x-4); \ x = 5$
(c) $y = (2x+3)^2; \ x = 0$
(d) $y = \dfrac{1}{x} - 1; \ x = 2$

7. The equation of a curve is $y = (x-3)(x+4)$. Find the gradient of the curve

(a) at the point where the curve crosses the y-axis
(b) at each of the points where the curve crosses the x-axis

8. If the equation of a curve is $y = 2x^2 - 3x - 2$ find

(a) the gradient at the point where $x = 0$
(b) the coordinates of the points where the curve crosses the x-axis
(c) the gradient at each of the points found in (b).

9. Find the coordinates of the point(s) on the curve $y = 3x^3 - x + 8$ at which the gradient is (a) 8 (b) 0

10. Find $\dfrac{dy}{dx}$ if

   (a) $y = x^4 - x^2$ (b) $y = (3x+4)^2$ (c) $y = \dfrac{x-3}{\sqrt{x}}$

11. Find the gradient of the tangent at the point where $x = 1$ on the curve $y = (2-x)^2$.

12. Find the coordinates of the point on the curve $y = x^2$ where the gradient of the tangent is $\frac{1}{4}$.

13. The equation of a curve is $s = 4t^2 + 5t$. Find the gradient of the tangent at each of the points where the curve crosses the t-axis.

14. Find the coordinates of the points on the curve $y = x^3 - 6x^2 + 12x + 2$ at which the tangent is parallel to the line $y = 3x$.

15. The curve $y = (x-2)(x-3)(x-4)$ cuts the x-axis at the points $P(2, 0)$, $Q(3, 0)$ and $R(4, 0)$. Prove that the tangents at P and R are parallel.

16. For a certain equation, $\dfrac{dy}{dx} = 2x + 1$. Which of the following could be the given equation?

   (a) $y = 2x^2 + x$ (b) $y = x^2 + x - 1$
   (c) $y = x^2 + 1$ (d) $y = x^2 + x$

17. Find the gradient of the curve with equation $y = 6x^2 - x$ at the point where $x = 1$ Find the equation of the tangent at this point. Where does this tangent meet the line $y = 2x$?

18. Find the equation of the tangent to the curve $y = 1 - x^2$ at the point where the curve crosses the positive x-axis.

19. Find the coordinates of the points on the curve $y = x^3 + 3x$ where the gradient is 15.

20. Find the coordinates of the points where the tangents to the curve $y = x^3 - 6x^2 + 12x + 2$ are parallel to the line $y = 3x$.

CHAPTER 14

STATIONARY POINTS

STATIONARY VALUES

Consider a function $f(x)$. The derived function, $f'(x)$, expresses the rate at which $f(x)$ increases with respect to x

At points where $f'(x)$ is positive, $f(x)$ is increasing as x increases, whereas if $f'(x)$ is negative then $f(x)$ is decreasing as x increases.

Now there may be points where $f'(x)$ is zero, i.e. $f(x)$ is momentarily neither increasing nor decreasing with respect to x

The value of $f(x)$ at such a point is called a *stationary value* of $f(x)$

i.e. $f'(x) = 0 \Rightarrow f(x)$ has a stationary value.

To look at this situation graphically we consider the curve with equation $y = f(x)$.

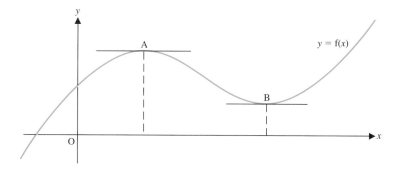

At A and B, $f(x)$, and therefore y, is neither increasing nor decreasing with respect to x. So the values of y at A and B are stationary values.

i.e. $\dfrac{dy}{dz} = 0 \Rightarrow y$ has a stationary value.

The point on a curve where y has a stationary value is called a *stationary point* and we see that, at any stationary point, the gradient of the tangent to the curve is zero, i.e. the tangent is parallel to the x-axis.

To sum up:

$$\frac{dy}{dx} \ [\text{or } f'(x)] \text{ is the rate of increase of } y \text{ w.r.t. } x$$

$$\text{at a stationary point} \begin{cases} y \ [\text{or } f(x)] \quad \text{has a stationary value} \\ dy/dx \quad [\text{or } f'(x)] \quad \text{is zero} \\ \text{the tangent is parallel to the } x\text{-axis.} \end{cases}$$

Example 14a

Find the stationary values of the function $x^3 - 4x^2 + 7$.

If $\qquad f(x) = x^3 - 4x^2 + 7$

then $\qquad f'(x) = 3x^2 - 8x$

At stationary points, $f'(x) = 0$, i.e. $3x^2 - 8x = 0$

$\Rightarrow \qquad x(3x - 8) = 0 \quad \Rightarrow \quad x = 0 \text{ and } x = \frac{8}{3}$

Therefore there are stationary points where $x = 0$ and $x = \frac{8}{3}$

When $x = 0$, $\qquad f(x) = 0 - 0 + 7 = 7$

When $x = \frac{8}{3}$, $\qquad f(x) = \left(\frac{8}{3}\right)^3 - 4\left(\frac{8}{3}\right)^2 + 7 = -2\frac{13}{27}$

Therefore the stationary values of $x^3 - 4x^2 - 5$ are 7 and $-2\frac{13}{27}$

EXERCISE 14a

Find the value(s) of x at which the following functions have stationary values.

1. $x^2 + 7$

2. $2x^2 - 3x - 2$

3. $x^3 - 4x^2 + 6$

4. $(2x - 3)^2$

5. $x^3 - 2x^2 + 11$

6. $x^3 - 3x - 5$

Find the value(s) of x for which y has a stationary value.

7. $y = x^2 - 8x + 1$

8. $y = x + 9/x$

9. $y = 2x^3 + x^2 - 8x + 1$

10. $y = 9x^3 - 25x$

11. $y = (2 - x)^2$

12. $y = 3x^3 - 12x + 19$

Find the coordinates of the stationary points on the following curves.

13. $y = \dfrac{x^2 + 9}{2x}$

14. $y = x^3 - 2x^2 + x - 7$

15. $y = (x-3)(x+2)$

16. $y = \left(2 - \dfrac{x}{2}\right)^2$

17. $y = \sqrt{x} + \dfrac{1}{\sqrt{x}}$

18. $y = 8 + \dfrac{x}{4} + \dfrac{4}{x}$

TURNING POINTS

In the immediate neighbourhood of a stationary point a curve can have any one of the shapes shown in the following diagram.

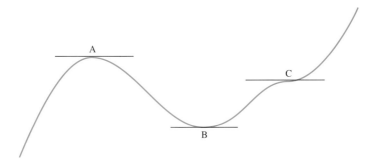

Moving through A from left to right we see that the curve is rising, then turns at A and begins to fall, i.e. the gradient changes from positive to zero at A and then becomes negative.

At A there is a *turning point*.

The value of y at A is called a *maximum value* and A is called a *maximum point*.

Moving through B from left to right the curve is falling, then turns at B and begins to rise, i.e. the gradient changes from negative to zero at B and then becomes positive.

At B there is a *turning point*.

The value of y at B is called a *minimum value* and B is called a *minimum point*.

The tangent is always horizontal at a turning point.

Note that a maximum value of y is *not necessarily the greatest value of y overall*. The terms maximum and minimum apply only to the behaviour of the curve in the neighbourhood of a stationary point.

At C the curve does not turn. The gradient goes from positive, to zero at C and then becomes positive again, i.e. the gradient does not change sign at C.

C is not a turning point but, because there is a change in the sense in which the curve is turning (from clockwise to anti-clockwise), C is called a *point of inflexion*.

INVESTIGATING THE NATURE OF STATIONARY POINTS

We already know how to locate stationary points on a curve and now examine several ways of distinguishing between the different types of stationary point.

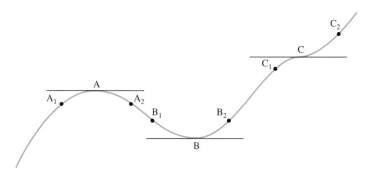

Method 1

This method compares the value of y at the stationary point with values of y at points on either side of, and near to, the stationary point.

For a maximum value, e.g. at A y at $A_1 < y$ at A

                                      y at $A_2 < y$ at A

For a minimum point, e.g. at B y at $B_1 > y$ at B

                                      y at $B_2 > y$ at B

For a point of inflexion, e.g. at C y at $C_1 < y$ at C

                                      y at $C_2 > y$ at C

Collecting these conclusions we have:

	Maximum	Minimum	Inflexion
y values on each side of the stationary point	both smaller	both larger	one larger and one smaller

Note that, between the points chosen on either side of the stationary point there must be no *other* stationary point, nor any discontinuity on the graph, e.g.

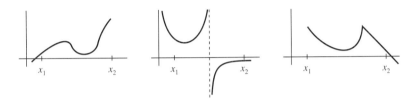

Method 2

This method examines the sign of the gradient, again at points close to, and on either side of, the stationary point where the gradient is zero.

For a maximum point, A, $\dfrac{dy}{dx}$ at A_1 is +ve, $\dfrac{dy}{dx}$ at A_2 is −ve

For a minimum point, B, $\dfrac{dy}{dx}$ at B_1 is −ve, $\dfrac{dy}{dx}$ at B_2 is +ve

For a point of inflexion, C, $\dfrac{dy}{dx}$ at C_1 is +ve, $\dfrac{dy}{dx}$ at C_2 is +ve

Collecting these conclusions we have:

Sign of $\dfrac{dy}{dx}$	Passing through maximum + 0 −	Passing through minimum − 0 +	Passing through point of inflexion + 0 + or − 0 −
Gradient of tangent	╱ ⌐ ╲	╲ _ ╱	╱ ─ ╲ or ─ ╲

Example 14b

Locate the stationary points on the curve $y = 4x^3 + 3x^2 - 6x - 1$ and determine the nature of each one.

$$y = 4x^3 + 3x^2 - 6x - 1 \quad \Rightarrow \quad \frac{dy}{dx} = 12x^2 + 6x - 6$$

At stationary points, $\dfrac{dy}{dx} = 0$

i.e. $\qquad 12x^2 + 6x - 6 = 0 \quad \Rightarrow \quad 6(2x-1)(x+1) = 0$

$\therefore$ there are stationary points where $x = \frac{1}{2}$ and $x = -1$

When $x = \frac{1}{2}$, $y = -2\frac{3}{4}$ and when $x = -1$, $y = 4$

i.e. the stationary points are $(\frac{1}{2}, -2\frac{3}{4})$ and $(-1, 4)$

Considering the sign of $\dfrac{dy}{dx}$ when $x = 0$ and at $x = 1$

(i.e. on either side of $x = \frac{1}{2}$) gives

$\left.\begin{array}{l} \dfrac{dy}{dx} = -6 \text{ when } x = 0 \\[2mm] \dfrac{dy}{dx} = 12 \text{ when } x = 1 \end{array}\right\} \quad \Rightarrow \quad (\frac{1}{2}, -2\frac{3}{4}) \text{ is a minimum point}$

Considering the sign of $\dfrac{dy}{dx}$ when $x = -\frac{3}{2}$ and at $x = -\frac{1}{2}$

(i.e. on either side of $x = -1$) gives

$\left.\begin{array}{l} \dfrac{dy}{dx} = 12 \text{ when } x = -\frac{3}{2} \\[2mm] \dfrac{dy}{dx} = -6 \text{ when } x = -\frac{1}{2} \end{array}\right\} \quad \Rightarrow \quad (-1, 4) \text{ is a maximum point}$

EXERCISE 14b

Find the stationary points on the following curves and distinguish between them.

1. $y = 2x - x^2$

2. $y = 3x - x^3$

3. $y = \dfrac{9}{x} + x$

4. $y = x^2(x - 5)$

5. $y = x^2$

6. $y = (2x - 5)^2$

7. $y = x^3$

8. $y = x^4$

9. $y = (2x + 1)(x - 3)$

10. $y = x^5 - 5x$

Find the stationary value(s) of each of the following functions and determine their character.

11. $x + \dfrac{1}{x}$

12. $3 - x + x^2$

13. $27x - x^3$

14. $x^2(3x^2 - 2x - 3)$

15. Use the information found from question 8, together with any other information you need, to sketch the curve $y = x^4$.

16. Repeat question 15 for the curve given in (a) question 10   (b) question 3.

PROBLEMS

Examples 14c

1. An open box is made from a square sheet of cardboard, with sides half a metre long, by cutting out a square from each corner, folding up the sides and joining the cut edges. Find the maximum capacity of the box.

The capacity of the box depends upon the unknown length of the side of the square cut from each corner so we denote this by x metres. The side of the cardboard sheet is $\frac{1}{2}$ m, so we know that $0 < x < \frac{1}{4}$

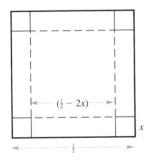 

Using metres throughout,

the base of the box is a square of side $(\frac{1}{2} - 2x)$
and the height of the box is x

∴   the capacity, C, of the box is given by

$$C = x(\tfrac{1}{2} - 2x)^2 = \tfrac{1}{4}x - 2x^2 + 4x^3 \quad \text{for } 0 < x < \tfrac{1}{4}$$

⇒ $\dfrac{dC}{dx} = \tfrac{1}{4} - 4x + 12x^2$

At a stationary value of C, $\dfrac{dC}{dx} = 0$

i.e. $\qquad 12x^2 - 4x + \frac{1}{4} = 0 \quad \Rightarrow \quad 48x^2 - 16x + 1 = 0$

$\qquad\qquad (4x - 1)(12x - 1) = 0 \quad \Rightarrow \quad x = \frac{1}{4} \text{ or } x = \frac{1}{12}$

there are stationary values of C when $x = \frac{1}{4}$ and when $x = \frac{1}{12}$

It is obvious that it is not possible to make a box if $x = \frac{1}{4}$ so it is tempting to *assume* that $x = \frac{1}{12}$ gives a maximum capacity. However we ought to check that this is the case by considering the signs of $\dfrac{dC}{dx}$ close to $x = \frac{1}{12}$ (i.e. $x \approx 0.08$).

x	0.07	0.09
$\dfrac{dC}{dx}$	$0.25 - 0.28 + 0.06 > 0$	$0.25 - 0.36 + 0.10 < 0$

Therefore C has a maximum value of $\frac{1}{12} \left(\frac{1}{2} - \frac{1}{6} \right)^2$, i.e. $\frac{1}{108}$

i.e. the maximum capacity of the box is $\frac{1}{108} \text{m}^3$

or, correct to 3 sf, 9260 cm^3

Alternatively the nature of the stationary point where $x = \frac{1}{12}$ can be investigated by using the sketch given by a graphics calculator for the curve.
$C = \frac{1}{4}x - 2x^2 + 4x^3$ and looking at the section for which $0 < x < \frac{1}{4}$, i.e.

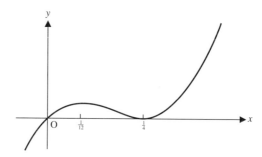

The sketch shows that there is a maximum point between $x = 0$ and $x = \frac{1}{4}$ so $x = \frac{1}{12}$ must give the maximum value of C.

The sketch also shows that there is a minimum point *on the curve* where $x = \frac{1}{4}$ but this is *not* a minimum value of the *capacity*, as a box cannot be made if $x = \frac{1}{4}$.

2. The function $ax^2 + bx + c$ has a gradient function $4x + 2$ and a stationary value of 1. Find the values of a, b and c

$$f(x) = ax^2 + bx + c \quad \Rightarrow \quad f'(x) = 2ax + b$$

But we know that $f'(x) = 4x + 2$

$\therefore \qquad 2ax + b$ is identical to $4x + 2$

i.e. $\qquad a = 2$ and $b = 2$

The stationary value of $f(x)$ occurs when $f'(x) = 0$

i.e. when $\qquad 4x + 2 = 0 \quad \Rightarrow \quad x = -\frac{1}{2}$

the stationary value of $f(x)$ is $2(-\frac{1}{2})^2 + 2(-\frac{1}{2}) + c = -\frac{1}{2} + c$

But the stationary value of $f(x)$ is also 1

$\therefore \qquad -\frac{1}{2} + c = 1 \quad \Rightarrow \quad c = \frac{3}{2}$

3. A cylinder has a radius r metres and a height h metres. The sum of the radius and height is 2 m. Find an expression for the volume, V cubic metres, of the cylinder in terms of r only. Hence find the maximum volume.

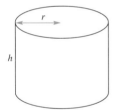

$$V = \pi r^2 h \text{ and } r + h = 2$$

$$\therefore \qquad V = \pi r^2 (2 - r) = \pi (2r^2 - r^3)$$

Now for maximum volume, $\dfrac{dV}{dr} = 0$,

i.e. $\qquad \pi(4r - 3r^2) = 0 \quad \Rightarrow \quad \pi r(4 - 3r) = 0$

Therefore there are stationary values of V when $r = 0$ and $r = \frac{4}{3}$. It is obvious that when $r = 0$, $V = 0$, and no cylinder exists, so we check that $r = \frac{4}{3}$ does give the maximum volume

r	1		$\frac{5}{3}$
$\dfrac{dV}{dr}$	$\pi \, (>0)$		$-\frac{5}{3}\pi \, (<0)$

Therefore the maximum value of V occurs when $r = \frac{4}{3}$ and is $\pi(\frac{4}{3})^2(2 - \frac{4}{3})$

i.e. the maximum volume is $\dfrac{32\pi}{27}\,\text{m}^3$

Note that the solution of this problem depends fundamentally on having an expression for V in terms of only one variable. This is true of *all* problems on stationary points so, if three or more variables are involved initially, some of them must be eliminated so that we have a basic relationship containing only two variables.

EXERCISE 14c

1. A farmer has an 80 m length of fencing. He wants to use it to form three sides of a rectangular enclosure against an existing fence which provides the fourth side. Find the maximum area that he can enclose and give its dimensions.

2. A large number of open cardboard boxes are to be made and each box must have a square base and a capacity of 4000 cm³. Find the dimensions of the box which contains the minimum area of cardboard.

3.

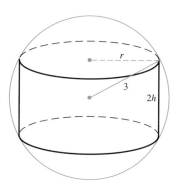

The diagram shows a cylinder cut from a solid sphere of radius 3 cm. Given that the cylinder has a height of $2h$, find its radius in terms of h. Hence show that the volume, V cubic metres, of the cylinder is given by

$$V = 2\pi h(9 - h^2)$$

Find the maximum volume of the cylinder as h varies.

4. A variable rectangle has a constant perimeter of 20 cm. Find the lengths of the sides when the area is maximum.

5. A variable rectangle has a constant area of 35 cm². Find the lengths of the sides when the perimeter is minimum.

6. The curve $y = ax^2 + bx + c$ crosses the y-axis at the point $(0, 3)$ and has a stationary point at $(1, 2)$. Find the values of a, b and c

7. The gradient of the tangent to the curve $y = px^2 - qx - r$ at the point $(1, -2)$ is 1. If the curve crosses the x-axis where $x = 2$, find the values of p, q and r. Find the other point of intersection with the x-axis and sketch the curve.

8. y is a quadratic function of x. The line $y = 2x$ is a tangent to the curve at the point $(3, 6)$. The turning point on the curve occurs where $x = -2$. Find the equation of the curve.

9. Locate the turning points on the curve $y = x(x^2 - 12)$, determine their nature and draw a rough sketch of the curve.

10. Find the stationary values of the function $x + 1/x$ and sketch the function.

11. If the perimeter of a rectangle is fixed in length, show that the area of the rectangle is greatest when it is square.

12. A door is in the shape of a rectangle surmounted by a semicircle whose diameter is equal to the width of the rectangle. If the perimeter of the door is 7 m, and the radius of the semicircle is r metres, express the height of the rectangle in terms of r. Show that the area of the door has a maximum value when the width is $14/(4 + \pi)$.

13. An open tank is constructed, with a square base and vertical sides, to hold 32 cubic metres of water. Find the dimensions of the tank if the area of sheet metal used to make it is to have a minimum value.

14. Triangle ABC has a right angle at C. The shape of the triangle can vary but the sides BC and CA have a fixed total length of 10 cm. Find the maximum area of the triangle.

*15. As part of an oil exploration exercise, a test bore hole is drilled. Instruments then measure the flow of oil (F) in the hole at various depths below the surface and present the results as a graph. This graph shows the results from one such test hole.

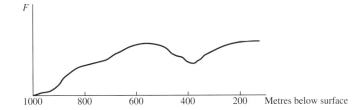

(a) Between what depths is the flow (i) increasing (ii) decreasing?

(b) What do you think is happening to the oil in the hole from about 400 to 600 metres below the surface?

(c) Sketch a graph showing, roughly, how the *rate* of flow varies with depth.

(d) Part of the hole is to be lined with impervious material to stop the leak of oil between the rock and the hole. Which section of the hole would you recommend for this treatment?

*16. A police car followed a lorry through a 2-mile section of road works on a motorway where there was a 50 m.p.h. speed limit. The on-board computer gave the following print out showing how the distance was covered.

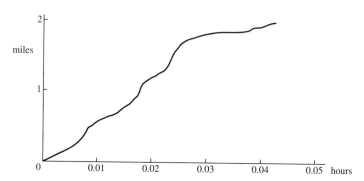

Use this graph to sketch another graph that shows, roughly, how the rate of increase of distance with respect to time (i.e. miles per hour) varied over this time.

Is there evidence that the lorry exceeded the speed limit at any time?

*17. In the last two questions, you needed to sketch a graph showing how the rate of change of the quantity on the vertical axis varied as the quantity on the horizontal axis increased, i.e. you produced a sketch of the gradient function. For each of the following curves, *sketch* a graph showing how dy/dx varies with x.

(a)   (b)

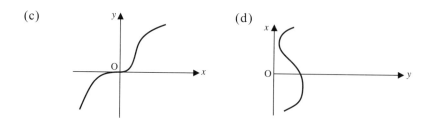

(c)   (d)

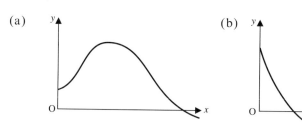

CHAPTER 15

TRIGONOMETRIC FUNCTIONS

THE GENERAL DEFINITION OF AN ANGLE

An angle is defined as a measure of the rotation of a line OP about a fixed point O. Taking Ox as the initial direction of OP, anticlockwise rotation describes a positive angle and clockwise rotation describes a negative angle. The rotation of OP is not limited to one revolution, so an angle can be as big as we choose to make it.

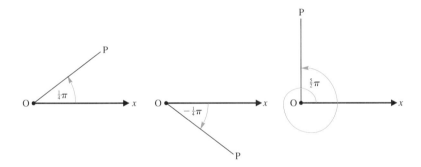

If θ is any angle, then θ can be measured

> either in degrees (one revolution $= 360°$)
>
> or   in radians (one revolution $= 2\pi$ radians)

and in either case we see that θ can take all real values.

THE TRIGONOMETRIC FUNCTIONS

Up to now the sine, cosine and tangent ratios have been defined only for acute and obtuse angles.

Now we know how an angle of *any* size is specified, we can define the sine, cosine and tangent of any angle θ.

If OP is drawn on x- and y-axes as shown and if, for all values of θ, the length of OP is r and the coordinates of P are (x, y), then the sine, cosine and tangent functions are defined as follows.

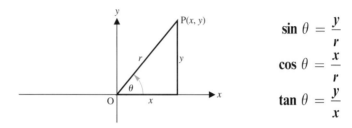

$$\sin \theta = \frac{y}{r}$$

$$\cos \theta = \frac{x}{r}$$

$$\tan \theta = \frac{y}{x}$$

We will now look at each of these functions in turn.

THE SINE FUNCTION

From the definition $f(\theta) = \sin \theta$, and measuring θ in radians, we can see that

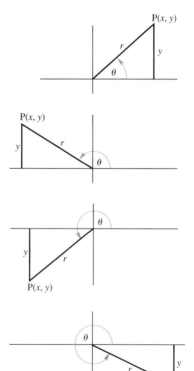

for $0 \leqslant \theta \leqslant \frac{1}{2}\pi$, OP is in the first quadrant; y is positive and increases in value from 0 to r as θ increases from 0 to $\frac{1}{2}\pi$. Now r is always positive, so $\sin \theta$ increases from 0 to 1

for $\frac{1}{2}\pi \leqslant \theta \leqslant \pi$, OP is in the second quadrant; again y is positive but decreases in value from r to 0, so $\sin \theta$ decreases from 1 to 0

for $\pi \leqslant \theta \leqslant \frac{3}{2}\pi$, OP is in the third quadrant; y is negative and decreases from 0 to $-r$, so $\sin \theta$ decreases from 0 to -1

for $\frac{3}{2}\pi \leqslant \theta \leqslant 2\pi$, OP is in the fourth quadrant; y is still negative but increases from $-r$ to 0, so $\sin \theta$ increases from -1 to 0

For $\theta \geqslant 2\pi$, the cycle repeats as OP travels round the quadrants again. For negative values of θ, OP rotates clockwise round the quadrants in the order 4th, 3rd, 2nd, 1st, etc. So $\sin\theta$ decreases from 0 to -1, then increases to 0 and on to 1 before decreasing to zero and repeating the pattern.

From this analysis we see that $\sin\theta$ is positive for $0 < \theta < \pi$ and negative when $\pi < \theta < 2\pi$

Further, $\sin\theta$ varies in value between -1 and 1 and the pattern repeats every revolution.

A plot of the graph of $f(\theta) = \sin\theta$ confirms these observations.

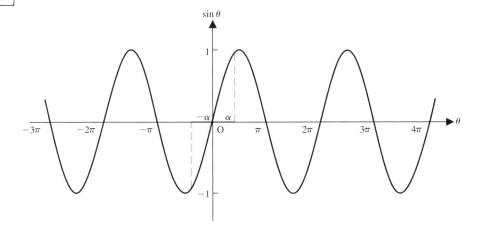

A graph of this shape is called, for obvious reasons, a *sine wave* and shows clearly the following characteristics of the sine function.

The curve is continuous (i.e. it has no breaks).

$$-1 \leqslant \sin\theta \leqslant 1$$

The shape of the curve from $\theta = 0$ to $\theta = 2\pi$ is repeated for each complete revolution.
Any function with a repetitive pattern is called *periodic* or *cyclic*.
The width of the repeating pattern, as measured on the horizontal scale, is called the *period*.

The period of the sine function is 2π

Other properties of the sine function shown by the graph are as follows.

$\sin \theta = 0$ when $\theta = n\pi$ where n is an integer.

The curve has rotational symmetry about the origin so the sine function is odd,

i.e. for any angle, α $\qquad$ $\mathbf{sin\,(-\alpha)\,=\,-sin\,\alpha}$

$$\text{e.g. } \sin(-30°) = -\sin(30°) = -\tfrac{1}{2}$$

$$\sin(-\tfrac{1}{2}\pi) = -\sin(\tfrac{1}{2}\pi) = -1$$

Also, if OP' is the reflection in the y-axis of OP, then OP' represents a rotation from Ox of $(180° - \theta°)$.

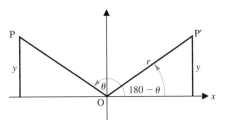

The values of y at P and P' are equal so $\sin\theta° = \sin(180° - \theta°)$ and the sine function is symmetrical about $\theta = 90°$.

Taking an enlarged section of the graph for $0 \leqslant \theta \leqslant 2\pi$, we find further relationships.

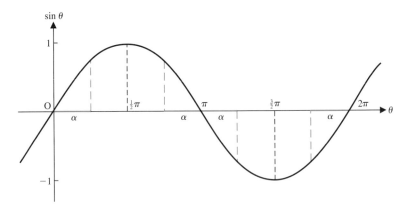

The curve is symmetrical about the line $\theta = \tfrac{1}{2}\pi$, so

$$\sin(\pi - \alpha) = \sin\alpha, \text{ e.g. } \sin130° = \sin(180° - 130°) = \sin50°$$

The curve has rotational symmetry about $\theta = \pi$, so

$$\sin(\pi + \alpha) = -\sin\alpha \text{ and } \sin(2\pi - \alpha) = -\sin\alpha$$

Examples 15a

1. Find the exact value of $\sin \frac{4}{3}\pi$.

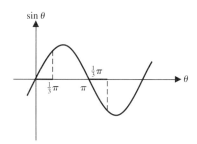

$$\sin \tfrac{4}{3}\pi = \sin\left(\pi + \tfrac{1}{3}\pi\right) = -\sin \tfrac{1}{3}\pi = -\frac{\sqrt{3}}{2}$$

2. Sketch the graph of $y = \sin\left(\theta - \tfrac{1}{4}\pi\right)$ for values of θ between 0 and 2π

Remember that the curve $y = f(x - a)$ is a translation of the curve $y = f(x)$ by a units in the positive direction of the x-axis.

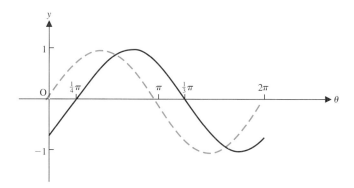

EXERCISE 15a

Find the exact value of

1. $\sin 120°$ 2. $\sin(-2\pi)$ 3. $\sin 300°$ 4. $\sin(-210°)$

5. $\sin(-60°)$ 6. $\sin \frac{13}{6}\pi$ 7. $\sin \frac{5}{4}\pi$ 8. $\sin\left(-\tfrac{2}{3}\pi\right)$

9. Write down all the values of θ between 0 and 6π for which $\sin\theta = 1$

10. Write down all the values of θ between 0 and -4π for which $\sin\theta = -1$

Express in terms of the sine of an acute angle.

11. $\sin 125°$ **12.** $\sin 290°$ **13.** $\sin(-120°)$ **14.** $\sin \frac{7}{6}\pi$

Sketch each of the following curves for values of θ in the range $0 \leqslant \theta \leqslant 3\pi$.

15. $y = \sin(\theta + \frac{1}{3}\pi)$ **16.** $y = -\sin\theta$ **17.** $y = \sin(-\theta)$

18. $y = 1 - \sin\theta$ **19.** $y = \sin(\pi - \theta)$ **20.** $y = \sin(\frac{1}{2}\pi - \theta)$

 Use a graphics calculator or computer for Questions 21 to 23 and set the range for θ as -2π to 4π.

21. On the same set of axes draw the graphs of $y = \sin\theta$, $y = 2\sin\theta$, and $y = 3\sin\theta$. What can you deduce about the relationship between the curves $y = \sin\theta$ and $y = a\sin\theta$?

22. On the same set of axes draw the curves $y = \sin\theta$ and $y = \sin 2\theta$.

23. On the same set of axes draw the curves $y = \sin\theta$ and $y = \sin 3\theta$. What can you deduce about the relationship between the two curves?

24. *Sketch* the curves (a) $y = \sin 4\theta$ (b) $y = 4\sin\theta$

One-way Stretches

Questions 21 to 24 in the last exercise show examples of one-way stretches. For example, the curve $y = 2\sin\theta$ is seen to be a one-way stretch of the curve $y = \sin\theta$ by a factor 2 parallel to the y-axis.

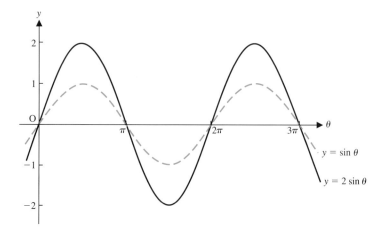

In general, if we compare points on the curves $y = f(x)$ and $y = af(x)$ with the same x-coordinate, then the y-coordinate of the point on $y = af(x)$ is a times the y-coordinate of the point on $y = f(x)$. Therefore

> **the curve $y = a$ or $f(x)$ is a one-way stretch of the curve $y = f(x)$ by a factor a parallel to the y-axis.**

Also, the curve $y = \sin 2\theta$ was seen to be a one-way stretch of the curve $y = \sin\theta$ by a factor $\frac{1}{2}$ parallel to the x-axis (or a one-way shrinkage by a factor 2).

Now consider points on the curves $y = f(x)$ and $y = f(ax)$ with the same y-coordinate. The x-coordinate on $y = f(ax)$ must be $\dfrac{1}{a}$ times the x-coordinate on $y = f(x)$. Therefore, in general,

> **the curve $y = f(ax)$ is a one-way stretch of the curve $y = f(x)$ by a factor $\dfrac{1}{a}$ parallel to the x-axis.**

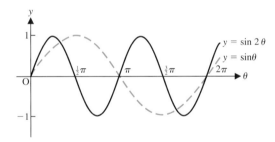

THE COSINE FUNCTION

For any position of P, $\cos\theta = \dfrac{x}{r}$

As OP rotates through the first quadrant, x decreases from r to 0 as θ increases, so $\cos\theta$ decreases from 1 to 0

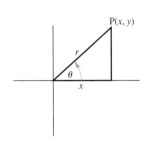

Similar observations of the behaviour of $\dfrac{x}{r}$ as OP continues to rotate show that

      in the second quadrant x decreases from 0 to $-r$,
      so $\cos\theta$ decreases from 0 to -1,
      in the third quadrant $\cos\theta$ increases from -1 to 0,
      and in the fourth quadrant $\cos\theta$ increases from 0 to 1

The cycle then repeats, and we get this graph of $f(\theta) = \cos\theta$

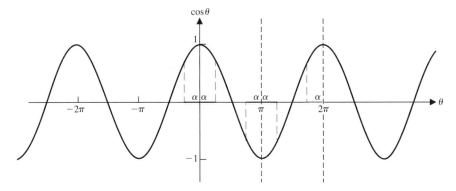

The characteristics of this graph are as follows.

The curve is continuous

$$-1 \leqslant \cos\theta \leqslant 1$$

It is periodic with a period of 2π

It is the same shape as the sine wave but is translated a distance $\frac{1}{2}\pi$ to the left. Such a translation of a sine wave is called a *phase shift*.

$$\cos\theta = 0 \quad\text{when}\quad \theta = \ldots -\tfrac{1}{2}\pi,\ \tfrac{1}{2}\pi,\ \tfrac{3}{2}\pi,\ \tfrac{5}{2}\pi,\ldots$$

The curve is symmetric about $\theta = 0$, so the cosine function is an even function and

$$\cos(-\alpha) = \cos\alpha$$

The curve has rotational symmetry about $\theta = \frac{1}{2}\pi$, so

$$\cos(\pi - \alpha) = -\cos\alpha$$

Further considerations of symmetry show that

$$\cos(\pi + \alpha) = -\cos\alpha \quad\text{and}\quad \cos(2\pi - \alpha) = \cos\alpha$$

Note that each of these properties can be proved from the definition of $\cos\theta$ with the help of a quadrant diagram but they are easier to see and remember from the graph.

EXERCISE 15b

1. Write in terms of the cosine of an acute angle
   (a) $\cos 123°$ (b) $\cos 250°$ (c) $\cos(-20°)$ (d) $\cos(-154°)$

2. Find the exact value of
   (a) $\cos 150°$ (b) $\cos \frac{3}{2}\pi$ (c) $\cos \frac{5}{4}\pi$ (d) $\cos 6\pi$
   (e) $\cos(-60°)$ (f) $\cos \frac{7}{3}\pi$ (g) $\cos(-\frac{5}{6}\pi)$ (h) $\cos 225°$

3. Sketch each of the following curves.
   (a) $y = \cos(\theta + \pi)$ (b) $y = \cos(\theta - \frac{1}{3}\pi)$ (c) $y = \cos(-\theta)$

4. Sketch the graph of $y = \cos(\theta - \frac{1}{4}\pi)$ for values of θ between $-\pi$ and π. Use the graph to find the values of θ in this range for which
   (a) $\cos(\theta - \frac{1}{4}\pi) = 1$ (b) $\cos(\theta - \frac{1}{4}\pi) = -1$
   (c) $\cos(\theta - \frac{1}{4}\pi) = 0$

5. On the same set of axes, sketch the graphs $y = \cos\theta$ and $y = 3\cos\theta$.

6. On the same set of axes, sketch the graphs $y = \cos\theta$ and $y = \cos 3\theta$.

7. Sketch the graph of $f(\theta) = \cos 4\theta$ for $0 \leqslant \theta \leqslant \pi$.
   Hence find the values of θ in this range for which $f(\theta) = 0$

8. Sketch the graph of $y = \cos(\frac{1}{2}\pi - \theta)$
   What relationship does this suggest between $\sin\theta$ and $\cos(\frac{1}{2}\pi - \theta)$?
   Is there a similar relationship between $\cos\theta$ and $\sin(\frac{1}{2}\pi - \theta)$?

9. Sketch the graph of $f(\theta) = \cos 2\theta$ for $0 \leqslant \theta \leqslant 2\pi$.
   Hence find the values of θ in this range for which
   (a) $\cos 2\theta = 0$ (b) $\cos 2\theta = -1$

10. On the same set of axes *sketch* the graphs of $y = 2\cos\theta$ and $y = \cos 2\theta$ for $0 \leqslant \theta \leqslant \pi$. Use your sketch to *estimate* the value of θ for which $2\cos\theta = \cos 2\theta$

11. The graph of $y = a + \cos\theta$ passes through the point $(\pi, 2)$. Find the value of a and the maximum value of y.

THE TANGENT FUNCTION

For any position of P, $\tan \theta = \dfrac{y}{x}$

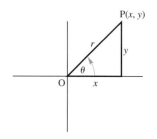

As OP rotates through the first quadrant, x decreases from r to 0 while y increases from 0 to r. This means that the fraction y/x increases from 0 to very large values indeed. In fact, as $\theta \to \frac{1}{2}\pi$, $\tan \theta \to \infty$

Looking at the behaviour of y/x in the other quadrants shows that in the second quadrant, $\tan \theta$ is negative and increases from $-\infty$ to 0, in the third quadrant, $\tan \theta$ is positive and increases from 0 to ∞, and in the fourth quadrant, $\tan \theta$ is negative and increases from $-\infty$ to 0   The cycle then repeats and we can draw the graph of $f(\theta) = \tan \theta$

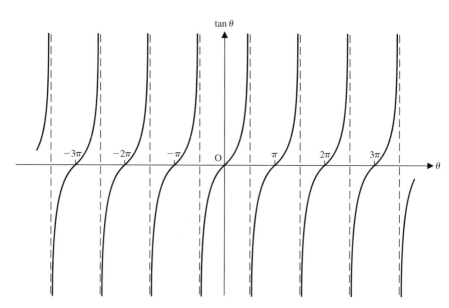

From the graph we can see that the characteristics of the tangent function are different from those of the sine and cosine functions in several respects.

It is not continuous, being *undefined* when $\theta = \ldots - \frac{1}{2}\pi,\ \frac{1}{2}\pi,\ \frac{2}{3}\pi, \ldots$

The range of values of $\tan\theta$ is unlimited.

It is periodic with a period of π (not 2π as in the other cases).

The graph has rotational symmetry about $\theta = 0$, so

$$\tan(-\alpha) = -\tan\alpha$$

The graph has rotational symmetry about $\theta = \frac{1}{2}\pi$, giving

$$\tan(\pi - \alpha) = -\tan\alpha$$

As the cycle repeats itself from $\theta = \pi$ to 2π, we have

$$\tan(\pi + \alpha) = \tan\alpha \quad\text{and}\quad \tan(2\pi - \alpha) = -\tan\alpha$$

Example 15c

Express $\tan\frac{11}{4}\pi$ as the tangent of an acute angle.

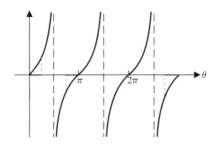

$$\tan(\tfrac{11}{4}\pi) = \tan(2\pi + \tfrac{3}{4}\pi) = \tan\tfrac{3}{4}\pi$$
$$= \tan(\pi - \tfrac{1}{4}\pi)$$
$$= -\tan\tfrac{1}{4}\pi$$

EXERCISE 15c

1. Find the exact value of

   (a) $\tan\frac{9}{4}\pi$ (b) $\tan 120°$ (c) $\tan -\frac{2}{3}\pi$ (d) $\tan\frac{7}{4}\pi$

2. Write in terms of the tangent of an acute angle

   (a) $\tan 220°$ (b) $\tan\frac{12}{7}\pi$ (c) $\tan 310°$ (d) $\tan -\frac{7}{5}\pi$

3. Sketch the graph of $y = \tan\theta$ for values of θ in the range 0 to 2π. From this sketch find the values of θ in this range for which

(a) $\tan\theta = 1$ (b) $\tan\theta = -1$ (c) $\tan\theta = 0$ (d) $\tan\theta = \infty$

4. Using the basic definitions of $\sin\theta$, $\cos\theta$ and $\tan\theta$, show that, for all values of θ,

$$\tan\theta = \frac{\sin\theta}{\cos\theta}$$

RELATIONSHIPS BETWEEN SIN θ, COS θ AND TAN θ

Because each trig ratio is a ratio of two of the three quantities x, y and r, we would expect to find several relationships between $\sin\theta$, $\cos\theta$ and $\tan\theta$. Some of these relationships will be investigated in a later module, but here is a summary of the results from the various exercises so far.

The graph of $\cos\theta$ shifted by $\frac{1}{2}\pi$ to the right gives the graph of $\sin\theta$.

Two angles which add up to $\frac{1}{2}\pi$ ($90°$) are called *complementary* angles,

so   **the sine of an angle is equal to the cosine of the complementary angle and vice-versa.**

Now $\sin\theta = \dfrac{y}{r}$, $\cos\theta = \dfrac{x}{r}$ and $\tan\theta = \dfrac{y}{x}$

$\therefore \quad \dfrac{\sin\theta}{\cos\theta} = \dfrac{y/r}{x/r} = \dfrac{y}{x} = \tan\theta$

i.e. for all values of θ, $\boldsymbol{\tan\theta = \dfrac{\sin\theta}{\cos\theta}}$

We have also seen that the sign of each trig ratio depends on the size of the angle, i.e. the quadrant in which P is. So we can summarise the sign of each ratio in a quadrant diagram:

sin + ve	all + ve
tan + ve	cos + ve

Examples 15d

1. Give all the values of x between 0 and $360°$ for which $\sin x = -0.3$

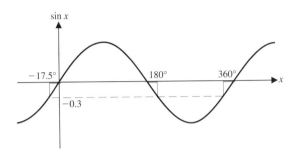

The value given for x by a calculator is $-17.5°$

From the graph, we see that, when $\sin x = -0.3$, the values of x in the specified range are $180° + 17.5°$ and $360° - 17.5°$

When $\sin x = -0.3$, $x = 197.5°$ and $342.5°$

(Note that when the range of values is given in degrees, the answer should also be given in degrees and the same applies for radians.)

2. Find the smallest positive value of θ for which $\cos \theta = 0.7$ and $\tan \theta$ is negative.

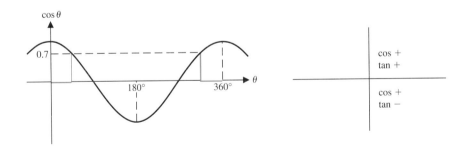

If $\cos \theta = 0.7$, the possible values of θ are $45.6°$, $314.4°$,...
Now $\tan \theta$ is positive if θ is in the first quadrant and negative if θ is in the fourth quadrant.

Therefore the required value of θ is $314.4°$

EXERCISE 15d

1. Within the range $-2\pi \leqslant \theta \leqslant 2\pi$, give all the values of θ for which

   (a) $\sin \theta = 0.4$ (b) $\cos \theta = -0.5$ (c) $\tan \theta = 1.2$

2. Within the range $0 \leqslant \theta \leqslant 720°$, give all the values of θ for which

   (a) $\tan \theta = -0.8$ (b) $\sin \theta = -0.2$ (c) $\cos \theta = 0.1$

3. Find the smallest angle (positive or negative) for which

   (a) $\cos \theta = 0.8$ and $\sin \theta \geqslant 0$

   (b) $\sin \theta = -0.6$ and $\tan \theta \leqslant 0$

   (c) $\tan \theta = \sin \frac{1}{6}\pi$

4. Using $\tan \theta = \dfrac{\sin \theta}{\cos \theta}$, show that the equation $\tan \theta = \sin \theta$ can be written as
   $\sin \theta (\cos \theta - 1) = 0$, provided that $\cos \theta \neq 0$
   Hence find the values of θ between 0 and 2π for which $\tan \theta = \sin \theta$.

5. Sketch the graph of $y = \sin 2\theta$. Use your sketch to help find the values of θ in the range $0 \leqslant \theta \leqslant 360°$ for which $\sin 2\theta = 0.4$

6. Sketch the graph of $y = \cos 3\theta$. Hence find the values of θ in the range $0 \leqslant \theta \leqslant 2\pi$ for which $\cos 3\theta = -1$

7. In the range $-360° \leqslant \theta \leqslant 360°$, sketch the graph of

   (a) $\sin (\theta - 30°)$ (b) $\sin (\theta + 30°)$

8. Sketch the graph of

   (a) $1 + \sin \theta$ (b) $1 - \cos \theta$ (c) $2 \sin \theta - 1$

9. (a)   Find the value of a for which the point $(\pi, 3)$ lies on the graph
          $y = a + \sin b\theta$

   (b)   If the point $\left(\dfrac{\pi}{2}, 3 \right)$ also lies on this graph, find a possible value for b.

EQUATIONS INVOLVING MULTIPLE ANGLES

Several simple equations have been solved in this chapter.

Now we will look at other methods for solving trig equations that involve ratios of a multiple of θ, for example

$$\cos 2\theta = \tfrac{1}{2} \quad \tan 3\theta = -2$$

Simple equations of this type can be solved by finding first the values of the multiple angle and then, by division, the corresponding values of θ

However it must be remembered that if values of θ are required in the range $\alpha \leqslant \theta \leqslant \beta$ then values of 2θ will be needed in double that range, i.e. $2\alpha \leqslant 2\theta \leqslant 2\beta$ and similarly for other multiples of θ

Examples 15e

1. Find the angles in the interval $\left[-\tfrac{1}{2}\pi, \tfrac{1}{2}\pi\right]$ which satisfy the equation
$\cos 2\theta = \tfrac{1}{2}$

As values of θ in the interval $\left[-\tfrac{1}{2}\pi, \tfrac{1}{2}\pi\right]$ are required, we need to find values of 2θ in the interval $[-\pi, \pi]$.

In the interval $[-\pi, \pi]$, the solutions of $\cos 2\theta = \tfrac{1}{2}$ are
$2\theta = \pm\tfrac{1}{3}\pi$

Hence $\theta = \pm\tfrac{1}{6}\pi$

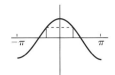

EQUATIONS INVOLVING COMPOUND ANGLES

When a compound angle appears in a trig equation such as

$$\cos\left(\theta - \tfrac{1}{4}\pi\right) = \tfrac{1}{2}$$

the equation can be solved by first finding values of the compound angle. The values of θ can then be found from a simple linear equation. If values of θ are required in the range $0 \leqslant \theta \leqslant \pi$ say,, then we must find the values of $\theta + \alpha$ in the interval $[0 + \alpha, \pi + \alpha]$

Examples 15e (continued)

2. Solve the equation $\cos(\theta - 20°) = -\frac{1}{2}$ for values of θ in the range $-180° \leqslant \theta \leqslant 180°$

As values of θ are required in the interval $[-180°, 180°]$, we need values of $(\theta - 20°)$ in the interval $[-200°, 160°]$.

In the interval $[-200°, 160°]$, the solutions of $\cos(\theta - 20°) = -\frac{1}{2}$ are $\theta - 20° = \pm 120°$

Hence

$\theta = 120° + 20°$ or $-120° + 20°$

$\quad = 140°$ or $-100°$

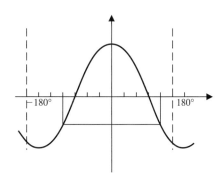

EXERCISE 15e

Find the values of θ in the range $0 \leqslant \theta \leqslant 360°$ which satisfy the following equations.

1. $\tan 2\theta = 1$

2. $\cos 3\theta = -0.5$

3. $\sin \frac{1}{2}\theta = -\frac{\sqrt{2}}{2}$

4. $\cos 5\theta = \frac{1}{2}$

5. $\tan \frac{1}{3}\theta = -\frac{1}{4}$

6. $\cos 2\theta = 0.63$

7. $\cos(\theta - 45°) = 0$

8. $\sin(\theta + 30°) = -1$

9. $\tan(\theta - 60°) = 0$

10. $\cos(\theta + 60°) = \frac{1}{2}$

Solve the following equations for values of θ in the range $-\pi \leqslant \theta \leqslant \pi$

11. $\cos(\theta + \frac{1}{4}\pi) = \frac{1}{2}$

12. $\tan(\theta - \frac{1}{3}\pi) = -1$

13. $\sin(\theta + \frac{1}{6}\pi) = \frac{1}{2}$

14. $\cos(\theta - \frac{1}{3}\pi) = -\frac{1}{2}$

Solve these equations for values of θ in the range $-180° \leqslant \theta \leqslant 180°$

15. $\tan 2\theta = 1.8$

16. $\sin 3\theta = 0.7$

17. $\cos \frac{1}{2}\theta = 0.85$

Solve these equations for values of θ in the range $0 \leqslant \theta \leqslant 2\pi$

$\tan 4\theta = -\sqrt{3}$

19. $\cos 5\theta = \frac{1}{2}$

20. $\tan \frac{1}{2}\theta = -1$

CHAPTER 16

NUMERICAL SOLUTION OF EQUATIONS

GRAPHICAL SOLUTIONS OF TRIG EQUATIONS

If you have worked through all the exercises in the last chapter, you will already have solved several simple trig equations with the help of sketch graphs. In this section we are going to look at more complicated equations whose solutions require accurate plots of graphs.

Consider the equation $\theta = 3 \sin \theta$

The values of θ for which $\theta = 3 \sin \theta$ can be found by plotting the graphs of $y = \theta$ and $y = 3 \sin \theta$ on the same axes and hence finding the values of θ at points of intersection.

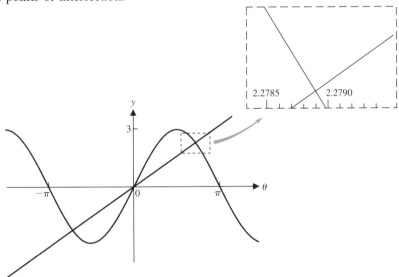

From the enlarged section of the graph $\theta = 2.2789 \, \text{rad}$

Therefore the three points of intersection occur where

$$\theta = -2.2789 \, \text{rad}, \quad 0, \quad 2.2789 \, \text{rad correct to 4 dp}$$

If these graphs are produced on a graphics calculator or on a computer using suitable software, then it is possible to zoom in on the points of intersection and get very accurate values for θ.   If the graphs are hand drawn, the accuracy of the results will depend on the patience and accuracy of the drawer! Plotting accurate graphs manually is tedious, so if you do not have either of the tools mentioned above, try just Question 1 in the following exercise and do not attempt to get answers correct to more than 2 sf.

EXERCISE 16a

1. Plot the graphs of $y = \theta$ and $y = 2\cos\theta$ for values of θ in the range $-\pi \leqslant \theta \leqslant \pi$. Hence find the values of θ for which $\theta = 2\cos\theta$

2. Repeat Question 1 using *sketch* graphs and measuring θ in degrees. If this was plotted accurately, would it give the same solutions as when the angle is measured in radians?

3. Measuring the angle in radians throughout, find graphically the values of θ for which

   (a) $2\theta = 4\sin\theta$ (b) $\sin\theta = \theta^2$ (c) $\cos\theta = \theta - 1$

USING CALCULATORS TO SOLVE EQUATIONS

Any equation which can be written in the form $f(x) = g(x)$ can be solved graphically provided that all constants have numerical values. For example, the equation $x^2 + 3x - 1 = 0$ can be solved graphically but the equation $x^2 - ax + 2 = 0$ can not, because a is not known.

Graphics calculators enable numerical solutions be found for any function that can be graphed on it. Using the zoom facility, it is possible to get these solutions accurate to as many significant figures as the calculator is capable of displaying.

Starting with the equation in the form $f(x) = g(x)$,

- either both $y = f(x)$ and $y = g(x)$ can be plotted, and the solution is given by the x-coordinate(s) of their point(s) of intersection,
- or the equation can be rearranged in the form $h(x) = 0$ so that just one curve, $y = h(x)$, needs to be plotted; the solutions are the values of x where the curve cuts the x-axis.

A pair of simultaneous equations in two unknowns, x and y, can also be solved graphically by rearranging the equations in the forms $y = f(x)$ and $y = g(x)$ and plotting their graphs. The solution(s) are the x *and* y coordinates of the points of intersection of the graphs.

There are also calculators which give solutions to quadratic equations, and to linear simultaneous equations, by simply entering the numerical values of the coefficients.

Omit the next exercise if you do not have access to a graphics calculator or to a computer with graph-drawing software.

EXERCISE 16b

Use a graphics calculator to solve the equations, giving all roots correct to 3 significant figures.

1. $4x^2 - 3x - 5 = 0$ 2. $x^3 = 5x$ 3. $x^4 - 3x + 1 = 0$

4. $2x - 2y = 7$ 5. $x + y = 7$ 6. $xy = 2$
    $5x + y = 8$ $x^2 + y = xy$ $x^2 + 2y = 5$

Set the range to $-\pi$, to π, and find the roots of these equations in that range.

7. $\cos x + \sin x = 1$ 8. $(\cos x)^2 - \sin x = 0$

9. $3\tan x = 2\sin x$ 10. $\sin x \cos x = 0.5$

LOCATING THE ROOTS OF THE EQUATION
$f(x) = 0$

A graphics calculator is a wonderful tool, but pencil and paper has its advantages – there are no batteries to run out for a start, and it is cheap. In the next section we look at a method for solving equations that does not rely on specialised calculators.

When the roots of an equation cannot be found exactly, we can find approximate solutions. The first step is to locate the roots roughly and this can be done graphically. It is sensible to rearrange the equation so that the function on each side of the equality is one that can be sketched easily.

Consider, for example, the equation $x^3 = x + 1$,

The roots of this equation are the values of x where the curve $y = x^3$ and the line $y = x + 1$ intersect.

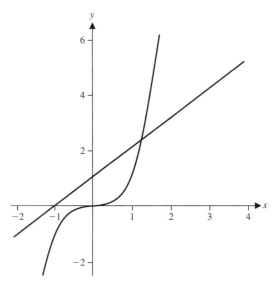

From the sketch we can see that there is one root between 1 and 2.

We now need a way to locate the roots more accurately.

Suppose that we have very roughly located the roots of an equation $f(x) = 0$

Now consider the curve $y = f(x)$

The roots of the equation $f(x) = 0$ are the values of x where this curve crosses the x-axis, e.g.

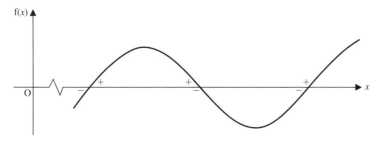

Each time that the curve crosses the x-axis, the sign of y changes. So

if one root only of the equation $f(x) = 0$ lies between x_1 and x_2, and if the curve $y = f(x)$ is unbroken between the points where $x = x_1$ and $x = x_2$, then $f(x_1)$ and $f(x_2)$ are opposite in sign.

The condition that the curve $y = f(x)$ must be unbroken between x_1 and x_2 is essential as we can see from the curve below.

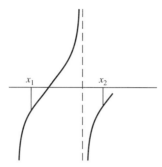

This curve crosses the x-axis between x_1 and x_2 but $f(x_1)$ and $f(x_2)$ have the same sign because the curve is broken between these values.

Returning to the equation $x^3 = x + 1$, we will now locate the larger root a little more precisely.

First we write the equation in the form $f(x) = 0$, i.e. $f(x) = x^3 - x - 1$ then we find where there is a change in the sign of $f(x)$

We know that there is a root between $x = 1$ and $x = 2$ so we will see if it lies between 1.0 and 1.5.

Using $f(x) = x^3 - x - 1$ gives $f(1) = 1^3 - 1 - 1 = -1$

$$\text{and } f(1.5) = (1.5)^3 - 1.5 - 1 = 0.875$$

Therefore the root of the equation lies between 1 and 1.5 and is likely to be nearer to 1.5 as $f(1.5)$ is nearer to zero than $f(1)$ is. This process can be repeated to narrow down the interval within which the root lies.

Example 16c

Find the turning points on the curve $y = x^3 - 2x^2 + x + 1$. Hence sketch the curve and use the sketch to show that the equation $x^3 - 2x^2 + x + 1 = 0$ has only one real root. Find two consecutive integers between which this root lies.

At turning points $\dfrac{dy}{dx} = 0$, i.e. $3x^2 - 4x + 1 = 0$

$$\Rightarrow \quad (3x - 1)(x - 1) = 0 \text{ so } x = \tfrac{1}{3} \text{ and } 1$$

When $x = \tfrac{1}{3}$, $y = \tfrac{31}{27}$ and when $x = 1$ $y = 1$

As the curve is a cubic, and as $y \to \infty$ as $x \to \infty$, we deduce that the curve has a maximum point at $\left(\tfrac{1}{3}, \tfrac{31}{27}\right)$ and a minimum point at $(1, 1)$

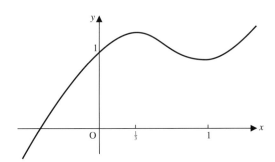

From the sketch, we see that the curve $y = x^3 - 2x^2 + x + 1$ crosses the x-axis at one point only, therefore

the equation $x^3 - 2x^2 + x + 1 = 0$ has only one real root.

Also from the sketch it appears that this root lies between $x = -1$ and $x = 0$

Using $f(x) = x^3 - 2x^2 + x + 1$ gives $f(-1) = -3$ and $f(0) = 1$

As $f(-1)$ and $f(0)$ are opposite in sign, the one real root of $x^3 - 2x^2 + x + 1 = 0$ lies between $x = -1$ and $x = 0$

EXERCISE 16c

1. Use sketch graphs to determine the number of real roots of each equation. (Some may have an infinite set of roots.)

(a) $x^3 = x$

(b) $\cos x = x^2 - 1$

(c) $1 + x^3 = \dfrac{1}{x}$

(d) $\sin x = \dfrac{1}{x}$

(e) $(x^2 - 4) = \dfrac{1}{x}$

(f) $x^4 = \dfrac{2}{x}$

2. For each equation in Question 1 with a finite number of roots, locate the root, or the larger root where there is more than one, within an interval of half a unit.

3. Find the turning points on the curve whose equation is $y = x^3 - 3x^2 + 1$. Hence sketch the curve and use your sketch to find the number of real roots of the equation $x^3 - 3x^2 + 1 = 0$

4. Using a method similar to that given in Question 3, or otherwise, determine the number of real roots of each equation.

(a) $x^4 - 3x^3 + 1 = 0$

(b) $x^3 - 24x + 1 = 0$

(c) $x^5 - 5x^2 + 4 = 0$

5. Given $\dfrac{1}{x^2} = \dfrac{1}{2}(x + 3)$, find the successive integers between which the smallest root lies.

CONSOLIDATION D

SUMMARY

Curves

A chord is a straight line joining two points on a curve.

A tangent to a curve is a line that touches the curve at one point, called the point of contact.

A normal to a curve is the line perpendicular to a tangent and through its point of contact.

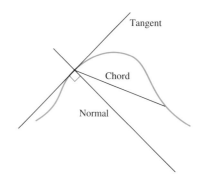

The gradient of a curve at a point on the curve is the gradient of the tangent at that point.

$y = af(x)$ is a one-way stretch of $y = f(x)$ by a factor, a, parallel to Oy

$y = f(ax)$ is a one-way stretch of $y = f(x)$ by a factor, $1/a$, parallel to Ox

Differentiation

Differentiation is the process of finding a general expression for the gradient of a curve at any point on the curve.

This general expression is called the gradient function, or the derived function or the derivative.

The derivative is denoted by $\dfrac{dy}{dx}$ or by $f'(x)$ where

$$\frac{dy}{dx} = \lim_{\delta x \to 0} \left[\frac{f(x + \delta x) - f(x)}{\delta x} \right]$$

When $y = x^n$, $\dfrac{dy}{dx} = nx^{n-1}$

When $y = ax^n$, $\dfrac{dy}{dx} = anx^{n-1}$

When $y = c$, $\dfrac{dy}{dx} = 0$

STATIONARY VALUES

A stationary value of $f(x)$ is its value where $f'(x) = 0$

The point on the curve $y = f(x)$ where $f(x)$ has a stationary value is called a stationary point.

At all stationary points, the tangents to the curve $y = f(x)$ are parallel to the x-axis.

TURNING POINTS

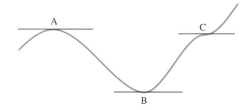

A, B and C are stationary points on the curve $y = f(x)$
The points A and B are called turning points.
The point C is called a point of inflexion.

At A, $f(x)$ has a maximum value and A is called a maximum point.
At B, $f(x)$ has a minimum value and B is called a minimum point.

There are two methods for distinguishing stationary points:

		Max	Min	Inflexion
1.	Find value of y on each side of stationary value	Both smaller	Both larger	One smaller One larger
2.	Find sign of $\dfrac{dy}{dx}$ on each side of stationary value	$+ \; 0 \; -$	$- \; 0 \; +$	$+ \; 0 \; +$ or $- \; 0 \; -$
	Gradient	$/ - \backslash$	$\backslash \_ /$	$/ - \backslash$ or $\backslash - \backslash$

Trigonometric Functions

The sine function, $f(x) = \sin x,$

is defined for all values of x

is periodic with a period 2π

has a maximum value of 1 when $x = (2n + \frac{1}{2})\pi$

and a minimum value of -1 when $x = (2n + \frac{3}{2})\pi$

is zero when $x = n\pi$

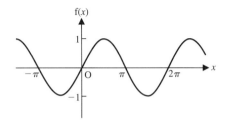

The cosine function, $f(x) = \cos x,$

is defined for all values of x

is periodic with a period 2π

has a maximum value of 1 when $x = 2n\pi$

and a minimum value of -1 when $x = (2n - 1)\pi$

is zero when $x = \frac{1}{2}(2n + 1)\pi$

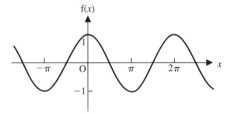

The tangent function, $y = \tan x,$

is undefined for some values of x

these values being all odd multiples of $\frac{1}{2}\pi$

is periodic with period π

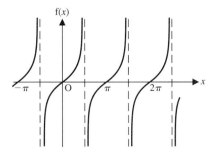

Transformations

It was shown earlier that

$y = f(x) + c$ is a translation of $y = f(x)$ by c units in the positive direction of the y-axis,

$y = f(x + c)$ is a translation of $y = f(x)$ by c units in the negative direction of the x-axis,

$y = -f(x)$ is the reflection of $y = f(x)$ in the x-axis,

$y = f(-x)$ is the reflection of $y = f(x)$ in the y-axis.

In addition, we now know that

$y = af(x)$ is a one-way stretch of $y = f(x)$ by a factor a parallel to the y-axis,

$y = f(ax)$ is a one-way stretch of $y = f(x)$ by a factor $1/a$ parallel to the x-axis.

Locating the Root of an Equation

The equation $f(x) = 0$ has a root between $x = a$ and $x = b$ if $f(a)$ and $f(b)$ are opposite in sign, provided that the curve $y = f(x)$ is unbroken between the points where $x = a$ and $x = b$.

MULTIPLE CHOICE EXERCISE D

TYPE I

1. The minimum value of $(x + 1)(x - 5)$ occurs where x is

A   1   **B**   5   **C**   2   **D**   0   **E** -2

2. The curve $y = \sin(x - 30°)$ could be

A

C

E
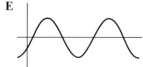

B

D
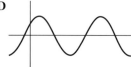

3. The function $x^3 - 12x + 5$ has a stationary value when

   A $x = \sqrt{6}$ **C** $x = 0$ **E** $x = 1$

   B $x = -2$ **D** $x = 4$

4. When $x = 1$ the function $x^3 - 3x^2 + 7$ is

   A   stationary   **C**   maximum   **E**   minimum

   B   increasing   **D**   decreasing

5. The rate of increase w.r.t. x of the function $x^2 - \dfrac{1}{x^2}$ is

   A $2x + \dfrac{2}{x^3}$ **C** $2x - \dfrac{2}{x^3}$ **E** $2x + \dfrac{1}{2x}$

   B $2x - \dfrac{2}{2x}$ **D** $2x + \dfrac{3}{x^3}$

6. The gradient function of $y = (x - 3)(x^2 + 2)$ is

   A $2x$ **C** $3x^2 - 6x + 2$ **E** $x^3 - 3x^2 + 2x - 6$

   B $2x - 3$ **D** $-3(2x + 2)$

7. The graph of the function $f(\theta) \equiv \cos 2\theta$ has a period

   A 2π **C** $\frac{1}{2}\pi$ **E**   none of these

   B π **D** -2π

TYPE II

8. When $\tan\theta = 1$, $\cos\theta = \sqrt{2}$

9. If $f(x) = x + \dfrac{1}{x}$, $\dfrac{d}{dx}f(x) = 1 - \dfrac{1}{x^2}$

10. When $y = x^3 - 4x + 5$, y is increasing when $x = 2$.

11. If $y = x^4$, x^4 has only one stationary value.

12. An angle θ is such that $\tan\theta = 1$ and $\cos\theta$ is negative. Hence $\sin\theta$ is positive.

13. When $f(\theta) = \cos\theta$, $f(\theta)$ is undefined when $\theta = \frac{1}{2}\pi$.

14. The graph of $f(\theta) = \cos\theta$ compared with the graph of $f(\theta) = \sin\theta$ is 90° to the left.

15. The function $f(\theta) = \cos\theta$ is such that $-1 \leqslant \theta \leqslant 1$.

16. $\sin\theta = 0$ when θ is a multiple of π.

17. A solution of the equation $\cos\theta = -1$ is $\theta = \pi$.

18. $y = \dfrac{1}{x^2} \quad \Rightarrow \quad \dfrac{dy}{dx} = \dfrac{-1}{2x}$

MISCELLANEOUS EXERCISE D

1. Given that $y = x^3 - x + 6$,

   (a) find $\dfrac{dy}{dx}$.

   On the curve representing y, P is the point where $x = -1$.

   (b) Calculate the y coordinate of the point P.

   (c) Calculate the value of $\dfrac{dy}{dx}$ at P.

   (d) Find the equation of the tangent at P.

   The tangent at the point Q is parallel to the tangent at P.

   (e) Find the coordinates of Q. (MEI)

2. The curve with equation $y = 2 + k \sin x$ passes through the point with coordinates $(\frac{1}{2}\pi, -2)$.
   Find

   (a) the value of k,

   (b) the greatest value of y,

   (c) the values of x in the interval $0 \leqslant x \leqslant 2\pi$ for which $y = 2 + \sqrt{2}$.
   (ULEAC)$_s$

3. Use differentiation to find the coordinates of the stationary points on the curve

   $$y = x + \dfrac{4}{x},$$

   and determine whether each stationary point is a maximum point or a minimum point.
   Find the set of values of x for which y increases as x increases. (UCLES)$_s$

4.

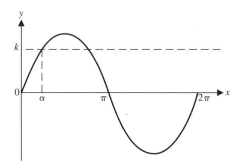

The diagram shows part of the graph of $y = \sin x$, where x is measured in radians, and the values α on the x-axis and k on the y-axis are such that $\sin \alpha = k$. Write down, in terms of α,

(a)   a value of x between $\frac{1}{2}\pi$ and π such that $\sin x = k$,

(b)   two values of x between 3π and 4π such that $\sin x = -k$.   (UCLES)ₛ

5. Given that $y = x^3 - 4x^2 + 5x - 2$, find $\dfrac{\mathrm{d}y}{\mathrm{d}x}$.

P is the point on the curve where $x = 3$.

(a)   Calculate the y coordinate of P.

(b)   Calculate the gradient at P.

(c)   Find the equation of the tangent at P.

(d)   Find the equation of the normal at P.   (MEI)ₛ

6. Sketch the graph of the curve with equation $y = x(1 - x)$. Determine the greatest and least values of y when $-1 \leqslant x \leqslant 1$ (UCLES)

7. The production cost per kilogram, C (in thousands of pounds), when x kilograms of a chemical are made is given by

$$C = 3x + \frac{100}{x}, x > 0$$

Find the value of x for which the cost is a minimum, and the minimum cost. Show, also, that this cost is a minimum rather than a maximum.   (OCSEB)

8. The profit, P, in thousands of pounds made by a company t months after its launch, where $0 \leqslant t \leqslant 12$, is given by

$$P = 2t^3 - 15t^2 + 24t + 2$$

Find $\dfrac{\mathrm{d}P}{\mathrm{d}t}$, and hence determine the range of values of t for which the profit is decreasing.   (ULEAC)

9. A lorry is to be driven 200 km at a steady speed of x km/hour.
   Petrol is consumed at the rate of $\left(3 + \dfrac{x^2}{240}\right)$ litres per hour.
   If petrol costs 50 p per litre and the driver earns £6 an hour, form an expression for the cost, £C, of the journey.
   Find the most economical speed and the total cost when this speed is maintained.

10. A cylinder, with an open top, is to be designed to hold 6000 cc of liquid.
    Show that the surface area is a minimum when the radius of the circular base
    is $10\left(\dfrac{6}{\pi}\right)^{1/3}$ cm.   (ULEAC)

11. (a)

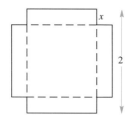 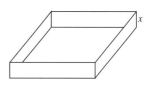

   The diagrams show how an open metal box may be formed. From a square thin sheet of metal measuring 2 units by 2 units, four equal squares of side x are removed; the projecting pieces are then folded upwards to make the sides of the box, which has depth x.
   Show that the volume V of the box is given by $V = 4x(1-x)^2$. Find the value of x for which V is a maximum, and find also the maximum value of V.

   (b) Sketch the graph of $y = 4x(1-x)^2$, and indicate on your sketch that part of the curve where x can represent the depth of a box as described in (a) above.   (OCSEB)

12. The rate of working, P watts, of an engine which is travelling at a speed $v\,\mathrm{m\,s}^{-1}$ is given by

$$P = 10v + \frac{4000}{v}, \quad v > 0$$

   Find the speed at which the rate of working is least.   (ULEAC)

13. Find the values of θ in the range $0 \leqslant \theta \leqslant 2\pi$ for which
    (a) $\sin\theta = 0.5$ (b) $\tan(\theta - \pi/3) = -1$

14. Show that the equation $x^3 - x^2 - 2 = 0$ has a root between 1 and 2.

15. Find the values of $\cos\theta$ for which $6\cos^2\theta - \cos\theta - 1 = 0$. Hence find the values of θ between 0 and 360° for which $6\cos^2\theta - \cos\theta - 1 = 0$.

16. The graph shows the curve whose equation is $y = x^4 - 4x - 1$.

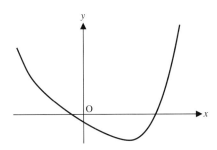

(a) Explain why this shows that the equation $x^4 - 4x - 1 = 0$ has two roots.

(b) Find two values of n such that $n < \alpha < n + 1$ where n is an integer and α is a root of the equation $x^4 - 4x - 1 = 0$.

17. Solve the equations for values of x in the range $-180° \leqslant x \leqslant 180°$

(a) $\sin 2x = 0.5$ (b) $\cos(x - 30°) = 0.5$

18. On the same set of axes, sketch the curves $y = 1 - \sin x$ and $y = \cos 2x$ for $0 \leqslant x \leqslant 360°$. Hence give the number of solutions of the equation $1 - \sin x = \cos 2x$ in the interval $[0, 360°]$.

19. Show that the equation $x^3 + 2x - 5 = 0$ has a root between 1 and 2. Determine which of the intervals $(1, 1.25)$, $(1.25, 1.5)$, $(1.5, 1.75)$, $(1.75, 2)$ this root lies in.

20. Show that there is a value of θ between $190°$ and $200°$ that satisfies the equation $8\cos^2 \theta = 7 - 2\sin \theta$ $(\cos^2 \theta = (\cos \theta)^2)$

CHAPTER 17

INTEGRATION

DIFFERENTIATION REVERSED

When x^2 is differentiated with respect to x the derivative is $2x$.

Conversely, if the derivative of an unknown function is $2x$ then it is clear that the unknown function could be x^2.

This process of finding a function from its derivative, which reverses the operation of differentiating, is called *integration*.

The Constant of Integration

As seen above, $2x$ is the derivative of x^2,
but it is also the derivative of $x^2 + 3$, $x^2 - 9$, and, in fact,
the derivative of $x^2 +$ any constant.

Therefore the result of integrating $2x$, which is called *the integral of 2x*, is not a unique function but is of the form

$$x^2 + K \text{ where } K \text{ is any constant}$$

K is called *the constant of integration*.

This is written

$$\int 2x \, dx = x^2 + K$$

where $\int \ldots dx$ means the integral of ... w.r.t. x

Integrating *any* function reverses the process of differentiating so, for any function $f(x)$ we have

$$\int \frac{d}{dx} f(x)\,dx = f(x) + K$$

e.g. because differentiating x^3 w.r.t. x gives $3x^2$ we have

$$\int 3x^2\,dx = x^3 + K$$

and it follows that $\int x^2\,dx = \frac{1}{3}x^3 + K$

Note that it is not necessary to write $\frac{1}{3}K$ in the second form, as K represents *any* constant in either expression.

In general, the derivative of x^{n+1} is $(n+1)x^n$ so

$$\int x^n\,dx = \frac{1}{(n+1)}x^{n+1} + K$$

i.e. **to integrate a power of** x,
increase **the power by 1 and** *divide* **by the new power.**

This rule can be used to integrate any power of x *except* -1.

EXERCISE 17a

Integrate with respect to x,

1. x^4
2. x^7
3. x^3

4. x^{11}
5. $\dfrac{1}{x^2}$
6. $\dfrac{1}{x^5}$

7. $x^{1/2}$
8. $x^{3/4}$
9. $\sqrt[3]{x}$

10. x^{-3}
11. $x^{-1/2}$
12. x

13. x^5
14. x^{-4}
15. $\sqrt[3]{x^2}$

16. Try using the rule to integrate $\dfrac{1}{x}$. Explain why it breaks down.

Integrating a Sum or Difference of Functions

We saw in Chapter 13 that a function can be differentiated term by term. Therefore, as integration reverses differentiation, integration also can be done term by term.

Example 17b

Find the integral of $x^7 + \dfrac{1}{x^2} - \sqrt{x}$

$$\int \left(x^7 + \frac{1}{x^2} - \sqrt{x} \right) dx = \int (x^7 + x^{-2} - x^{1/2}) \, dx$$

$$= \int x^7 \, dx + \int x^{-2} \, dx - \int x^{1/2} \, dx$$

$$= \frac{1}{8}x^8 + \frac{1}{-1}x^{-1} - \frac{1}{\frac{3}{2}}x^{3/2} + K$$

$$= \frac{1}{8}x^8 - \frac{1}{x} - \frac{2}{3}x^{3/2} + K$$

EXERCISE 17b

Integrate with respect to x.

1. $x^5 + \sqrt{x}$ 2. $\dfrac{1}{x^5} - x^2$ 3. $\sqrt[4]{x} + x^4$ 4. $x^{-3} - x^3$

5. $\dfrac{1}{x^{5/2}} + x^{2/5}$ 6. $x^{-1/2} + x^{1/2}$ 7. $x - \dfrac{1}{x^2}$ 8. $\dfrac{1}{\sqrt[3]{x}} + \sqrt{(x^3)}$

USING INTEGRATION TO FIND AN AREA

The area shown in the diagram is bounded by the curve $y = f(x)$, the x-axis and the lines $x = a$ and $x = b$.

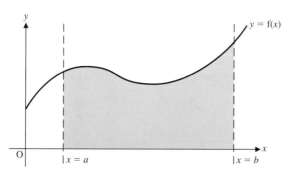

There are several elementary ways in which this area can be estimated, e.g. by counting squares on graph paper. A better method is to divide the area into thin vertical strips and treat each strip, or *element*, as being approximately rectangular.

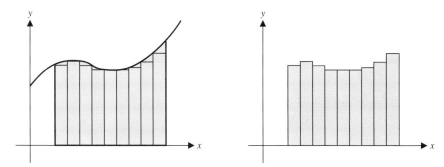

The sum of the areas of the rectangular strips then gives an approximate value for the required area. The thinner the strips are, the better is the approximation.

Note that every strip has one end on the x-axis, one end on the curve and two vertical sides, i.e., they all have the same type of boundaries.

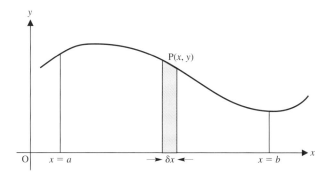

Now, considering a typical element bounded on the left by the ordinate through a general point $P(x, y)$, we see that

> **the width of the element represents a small increase in the value of x and so can be called δx**

Also, if A represents the part of the area up to the ordinate through P, then

> **the area of the element represents a small increase in the value of A and so can be called δA**

The shape of a typical strip is approximately a rectangle of height y and width δx.

Therefore, for any element

$$\delta A \simeq y\,\delta x \qquad\qquad\qquad\qquad [1]$$

The required area can now be found by adding the areas of all the strips from $x = a$ to $x = b$.

The notation for a summation of this kind is $\displaystyle\sum_{x=a}^{x=b} \delta A$

so,   total area $= \displaystyle\sum_{x=a}^{x=b} \delta A$

$\Rightarrow$ total area $\simeq \displaystyle\sum_{x=a}^{x=b} y\,\delta x$

As δx gets smaller the accuracy of the results increases until, in the limiting case,

$$\textbf{total area} = \lim_{\delta x \to 0} \sum_{x=a}^{x=b} y\,\delta x$$

Equation [1] above can also be written in the alternative form

$$\frac{\delta A}{\delta x} \simeq y$$

This form too becomes more accurate as δx gets smaller giving, in the limiting case,

$$\lim_{\delta x \to 0} \frac{\delta A}{\delta x} = y$$

But $\displaystyle\lim_{\delta x \to 0} \frac{\delta A}{\delta x}$ is $\dfrac{\mathrm{d}A}{\mathrm{d}x}$ so $\dfrac{\mathrm{d}A}{\mathrm{d}x} = y$

Hence $A = \displaystyle\int y\,\mathrm{d}x$

The boundary values of x defining the total area are $x = a$ and $x = b$ and we indicate this by writing

$$\textbf{total area} = \int_{a}^{b} y\,\mathrm{d}x$$

The total area can therefore be found in two ways, either as the limit of a sum or by integration

i.e.
$$\lim_{\delta x \to 0} \sum_{x=a}^{x=b} y\,\delta x = \int_a^b y\,dx$$

and we conclude that **integration is a process of summation.**

At this stage we will use integration only to find areas bounded by straight lines and a curve, but first we must investigate the meaning of $\int_a^b y\,dx$

DEFINITE INTEGRATION

Suppose that we wish to find the area bounded by the x-axis, the lines $x = a$ and $x = b$ and the curve $y = 3x^2$

Using the method above we find that $A = \int 3x^2\,dx$

i.e. $A = x^3 + K$

From this area function we can find the value of A corresponding to a particular value of x.

Hence using $x = a$ gives $\quad A_a = a^3 + K$

and using $x = b$ gives $\quad A_b = b^3 + K$

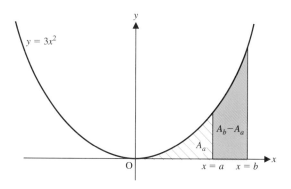

Then the area between $x = a$ and $x = b$ is given by $A_b - A_a$

where $A_b - A_a = (b^3 + K) - (a^3 + K) = b^3 - a^3$

Now $A_b - A_a$ is referred to as

the definite integral from a to b of $3x^2$

and is denoted by $$\int_a^b 3x^2 \, dx$$

i.e. $$\int_a^b 3x^2 \, dx = (x^3)_{x=b} - (x^3)_{x=a}$$

The RHS of this equation is usually written in the form $[x^3]_a^b$ where a and b are called the *boundary values* or *limits of integration*: b is the *upper limit* and a is the *lower limit*.

Whenever a definite integral is calculated, the constant of integration disappears.

Note. A definite integral can be found in this way only if the function to be integrated is defined for every value of x from a to b, e.g.

$\int_{-1}^{1} \dfrac{1}{x^2} \, dx$ cannot be found directly as $\dfrac{1}{x^2}$ is undefined when $x = 0$

Examples 17c

1. Find the value of $\displaystyle\int_0^9 x^{3/2} \, dx$

$$\int_0^9 x^{3/2} \, dx = \left[\tfrac{2}{5} x^{5/2} \right]_0^9$$
$$= \tfrac{2}{5} \{ (9)^{5/2} - 0 \}$$
$$= 97\tfrac{1}{5}$$

2. Evaluate $\displaystyle\int_1^4 \dfrac{\sqrt{x+1}}{\sqrt{x}}$

$$\int_1^4 \frac{\sqrt{x}+1}{\sqrt{x}} \, dx \equiv \int_1^4 (1 + x^{-1/2}) \, dx$$
$$= \left[x + 2x^{1/2} \right]_1^4$$
$$= \{4 + 4\} - \{1 + 2\}$$
$$= 5$$

EXERCISE 17c

Evaluate each of the following definite integrals.

1. $\int_0^1 x^4 \, dx$

2. $\int_1^2 x^{-2} \, dx$

3. $\int_1^4 x^{1/2} \, dx$

4. $\int_1^2 \frac{1}{x^5} \, dx$

5. $\int_4^9 \frac{1}{\sqrt{x}} \, dx$

6. $\int_0^1 (2x + 3)^2 \, dx$

7. $\int_0^2 x^3 \, dx$

8. $\int_1^2 \sqrt{x^5} \, dx$

9. $\int_2^4 (x^2 + 4) \, dx$

10. $\int_2^3 2x(x + 1) \, dx$

11. $\int_0^3 (x^2 + 2x - 1) \, dx$

12. $\int_0^2 (x^3 - 3x) \, dx$

13. $\int_{-1}^0 x(x - 1)(x - 2) \, dx$

14. $\int_{-1}^1 \frac{1}{x}(x^2 - x) \, dx$

15. $\int_{-1}^0 (x + 1)^2 \, dx$

16. $\int_1^2 \frac{(x + 1)^2}{\sqrt{x}} \, dx$

FINDING AREA BY DEFINITE INTEGRATION

As we have seen, the area bounded by a curve $y = f(x)$, the lines $x = a$, $x = b$, and the x-axis, can be found from the definite integral

$$\int_a^b f(x) \, dx$$

It is recommended, however, that this is not regarded as a *formula* but that the required area is first considered as the summation of the areas of elements, a typical element being shown in a diagram.

Example 17d

Find the area in the first quadrant bounded by the x and y axes and the curve $y = 1 - x^2$

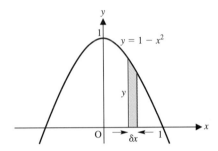

The required area starts at the y-axis, i.e. at $x = 0$ and ends where the curve crosses the x-axis, i.e. where $x = 1$. So it is given by

$$\lim_{\delta x \to 0} \sum_{x=0}^{x=1} y\,\delta x = \int_0^1 (1 - x^2)\,dx = \left[x - \frac{x^3}{3} \right]_0^1$$

$$= (1 - \tfrac{1}{3}) - (0 - 0) = \tfrac{2}{3}$$

The required area is $\tfrac{2}{3}$ of a square unit.

EXERCISE 17d

In each question find the area within the given boundaries.

1. The x-axis, the curve $y = x^2 + 3$ and the lines $x = 1$, $x = 2$.

2. The curve $y = \sqrt{x}$, the x-axis and the lines $x = 4$, $x = 9$.

3. The x-axis, the lines $x = -1$, $x = 1$, and the curve $y = x^2 + 1$.

4. The curve $y = x^2 + x$, the x-axis and the line $x = 3$.

5. The positive x- and y-axes and the curve $y = 4 - x^2$.

6. The lines $x = 2$, $x = 4$, the x-axis and the curve $y = x^3$.

7. The curve $y = 4 - x^2$, the positive y-axis and the negative x-axis.

8. The x-axis, the lines $x = 1$ and $x = 2$, and the curve $y = \frac{1}{2}x^3 + 2x$.

9. The x-axis and the lines $x = 1$, $x = 5$, and $y = 2x$.
   Check the result by sketching the required area and finding it by mensuration.

The Meaning of a Negative Result

Consider the area bounded by $y = 4x^3$ and the x-axis if the other boundaries are the lines

(a) $x = -2$ and $x = -1$ (b) $x = 1$ and $x = 2$

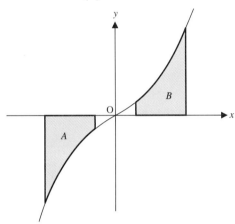

This curve is symmetrical about the origin so the two shaded areas are equal.

(a) Considering A

$$\lim_{\delta x \to 0} \sum_{x=-2}^{x=-1} y\,\delta x = \int_{-2}^{-1} y\,dx$$

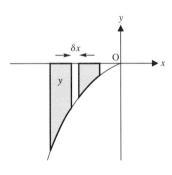

$$= \int_{-2}^{-1} 4x^3\,dx$$

$$= \left[x^4 \right]_{-2}^{-1}$$

$$= 1 - 16 = -15$$

(b) Considering B

$$\lim_{\delta x \to 0} \sum_{x=1}^{x=2} y\,\delta x = \int_{1}^{2} y\,dx$$

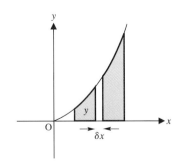

$$= \int_{1}^{2} 4x^3\,dx$$

$$= \left[x^4 \right]_{1}^{2}$$

$$= 16 - 1 = 15$$

So we see that, while the magnitudes of the two areas are equal, the result for the area of A, which is below the x-axis, is negative. This is explained by the fact that the length of a strip in A was taken as y, which is negative for the part of the curve bounding A.

Note. Care must be taken with problems involving a curve that crosses the x-axis between the boundary values.

Example 17e

Find the area enclosed between the curve $y = x(x-1)(x-2)$ and the x-axis.

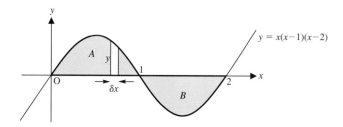

The area enclosed between the curve and the x-axis is the sum of the areas A and B.

For A we use
$$\int_0^1 y\,dx = \int_0^1 (x^3 - 3x^2 + 2x)\,dx$$

$$= \left[\frac{x^4}{4} - x^3 + x^2\right]_0^1$$

$$= \tfrac{1}{4}$$

For B we use
$$\int_1^2 (x^3 - 3x^2 + 2x)\,dx = \left[\frac{x^4}{4} - x^3 + x^2\right]_1^2$$

$$= (4 - 8 + 4) - (\tfrac{1}{4} - 1 + 1)$$

$$= -\tfrac{1}{4}$$

The minus sign refers only to the *position* of area B relative to the x-axis.
The actual area is $\tfrac{1}{4}$ of a square unit.

So the total shaded area is $\tfrac{1}{4} + \tfrac{1}{4} = \tfrac{1}{2}$ of a square unit.

1. Find the area enclosed by the curve $y = x^2$, the x-axis and the lines $x = 1$ $x = 2$.

2. Find the area bounded by the x- and y-axes, the lines $x = 1$ and part of the curve $y = x^3 + 2$.

3. Find the area of the region of the xy-plane which is enclosed by the x- and y-axes, the line $x = 3$ and part of the curve with equation $y = x^2 + 3x + 2$.

4. Find the area bounded by the curve $y = 1 - x^3$, the x-axis and the lines $x = 2$, $x = 3$.

5. The boundaries of a region in the xy plane are the lines $x = 1$, $x = 2$, the x-axis and part of the curve $y = x(3 - x)$. Find the area of this region.

6. Find the area below the x-axis and above the curve $y = x^2 - 1$.

7. Sketch the curve with equation $y = (x - 1)^2$. Calculate the area of the region enclosed by this curve and the x- and y-axes.

8. Sketch the curve $y = x(x^2 - 1)$, showing where it crosses the x-axis. Find

   (a) the area enclosed above the x-axis and below the curve
   (b) the area enclosed below the x-axis and above the curve
   (c) the total area between the curve and the x-axis.

9. Repeat Question 8 for the curve $y = x(4 - x^2)$.

10. Evaluate

   (a) $\int_0^2 (x - 2)\,dx$ (b) $\int_2^4 (x - 2)\,dx$ (c) $\int_0^4 (x - 2)\,dx$

   Interpret your results by means of a sketch.

USING HORIZONTAL ELEMENTS

Suppose that area between the curve $x = y(4 - y)$ and the y-axis is required.

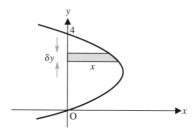

The curve crosses the y-axis where $y = 0$ and $y = 4$ as shown.

A vertical element is not suitable in this case because it has *both ends on the curve* and its length is therefore not easily found.

However it is easy to find the approximate area of a *horizontal* strip, by treating it as a rectangle with length x and width δy

i.e. area of element $\approx x \, \delta y$ and the required area is therefore given by

$$\lim_{\delta y \to 0} \sum_{y=0}^{y=4} x \, \delta y = \int_0^4 x \, dy = \int_0^4 y(y - 4) \, dy$$

EXERCISE 17f

1. Evaluate $\displaystyle\int_3^6 (y^2 - y) \, dy$

In Questions 2 to 5 find the area specified by the given boundaries.

2. The curve $x = y^2$, the y-axis and the lines $y = 1$, $y = 2$

3. The y-axis, the curve $x = \sqrt{y}$ and the line $y = 4$

4. The y-axis and the curve $x = 9 - y^2$

5. The y-axis and the curve $x = (y - 2)(y + 1)$

6. If $y = x^2$, show by means of sketch graphs and *not* by evaluating the integrals, that $\displaystyle\int_0^1 y \, dx = 1 - \int_0^1 x \, dy$

MIXED EXERCISE 17

Integrate with respect to x

1. $x^2 - 1/x^2$

2. $(3x + 7)^2$

3. $\sqrt{x} + 1/\sqrt{x}$

4. $\dfrac{2x - 1}{\sqrt{x}}$

5. $x^3 - \dfrac{1}{x^3}$

6. $\dfrac{x^2 - 1}{\sqrt{x}}$

Evaluate

7. $\displaystyle\int_{5}^{6} (6 - x)^2 \, dx$

8. $\displaystyle\int_{0}^{4} \dfrac{\sqrt{y} + 2}{\sqrt{y}} \, dy$

9. $\displaystyle\int_{1}^{32} \left(\sqrt[5]{x} - \dfrac{1}{\sqrt[5]{x}} \right) dx$

Find the areas specified in Questions 10 to 15

10. Between the curve $y = 6 + x - x^2$ and the x-axis.

11. Bounded by part of the curve $y = x^2 + 2$, the x-axis and the lines $x = 1$, $x = 4$.

12. Bounded by the x- and y-axes and the curve $y = 1 - x^3$.

13. Bounded by the y-axis, the lines $y = 1$, $y = 3$ and part of the curve $3y^2 = 2x - 3$.

14. Bounded by the curve $x = y^2 - 4$ and the y-axis.

15. The *total* area between the curve $y = (x - 1)(x - 2)(x - 3)$ and the x-axis.

16. (a) Sketch the curve with equation $y = (x + 2)(x - 5)$ marking the x-coordinates of the points where the curve crosses the x-axis.

    (b) Evaluate $\displaystyle\int_{-2}^{5} (x^2 - 3x - 10) \, dx$

    (c) Write down the value of the area between the curve and the x-axis.

17. (a) Sketch the curve $y = (x + 1)(x - 1)(x - 2)$.

    (b) Evaluate $\displaystyle\int_{-1}^{2} (x + 1)(x - 1)(x - 2) \, dx$

    (c) Find the area enclosed by the curve and the x-axis.

*18. When $\dfrac{dy}{dx} = f'(x)$ and $f'(x)$ is known, we cannot always find y as a function of x. This may be because we do not have the knowledge to do so, or it may be impossible.

For example, if $\dfrac{dy}{dx} = \dfrac{\sin x}{x}$, finding y as a function of x

$\left(\text{i.e. finding } \displaystyle\int \dfrac{\sin x}{x}\, dx \right)$, is beyond the scope of this book. It is however

possible to build up a picture of the shape of the graph of $y = \displaystyle\int \dfrac{\sin x}{x}\, dx$.

The diagram shows the graph of $\dfrac{dy}{dx} = \dfrac{\sin x}{x}$.

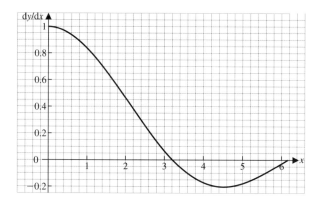

(a) A_1 is the area between the curve, the x-axis and the lines $x = 0$ and $x = 1$. Express A_1 in terms of a definite integral and find an approximate value for A_1 by counting squares.

(b) If y_0 is the value of y when $x = 0$, and y_1 is the value of y when $x = 1$, interpret the value of A_1 in terms of y_0 and y_1

(c) By finding approximate values for the area between the curve and various values of x, complete this table, given that $y_0 = 0.9$.

x	0	1	2	3	4	5	6
y							

(d) Plot the values of y against corresponding values of x and hence show, approximately, the graph of $y = f(x)$.

*19. The rate of inflation measures the rate at which the cost of a 'basket' of goods and services changes with time. The graph shows the rate of inflation over a six-year period.

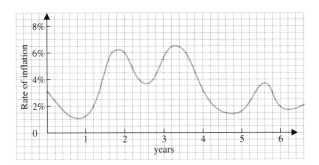

Given that the 'basket' cost £100 at the start of this six-year period, use the graph to show, approximately, how the cost of the 'basket' varied over the six years.

*20.

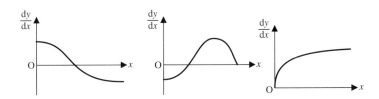

Each of the graphs shows dy/dx plotted against x. Sketch a graph showing y as a function of x, given that this curve goes through the origin.

CHAPTER 18

NUMBER SERIES

SEQUENCES

Consider the following sets of numbers,

     2, 4, 6, 8, 10, ...

     1, 2, 4, 8, 16, ...

     4, 9, 16, 25, 36, ...

Each set of numbers, in the order given, has a pattern and there is an obvious rule for obtaining the next number and as many subsequent numbers as we wish to find.

Such sets are called *sequences* and each member of the set is a term of the sequence.

The terms in a sequence are denoted by $u_1, u_2, \ldots, u_r, \ldots$ where u_r is the rth term.

Defining a Sequence

Consider the sequence 1, 2, 4, 8, 16, ...

Each term is a power of 2 so we can write the sequence as

     $2^0, 2^1, 2^2, 2^3, 2^4, \ldots$

All the terms are of the form 2^r, so 2^r is a general term.
however, 2^r is *not* the rth term, as $u_1 = 2^0$, $u_2 = 2^1$, $u_3 = 2^2, \ldots$

Now we can see that $u_r = 2^{r-1}$ and we can use this to find any term of the sequence,

e.g. the ninth term, u_9, is given by $u_9 = 2^{9-1} = 256$

Hence the rule $u_r = 2^{r-1}$ for $r = 1, 2, 3, \ldots$ enables the whole sequence to be generated and so defines it completely.

However it is not always possible, or desirable, to define a sequence by giving u_r in terms of r.

Consider the sequence 2, 4, 6, 8, 10,...

The obvious way to describe this sequence is 'starting with 2, add 2 to each term to get the next term,

or, using symbolic notation, $u_1 = 2$, $u_{r+1} = u_r + 2$

This is also a definition of the sequence since it can be generated as follows.

$$u_1 = 2, \ u_2 = u_1 + 2 = 4, \ u_3 = u_2 + 2 = 6, \ \text{and so on.}$$

When a sequence is generated from one known term together with a relationship between consecutive terms, the process is called *iteration* and the sequence is said to be defined iteratively. The relationship between u_r and u_{r+1} is called a *recurrence* relation.

Example 18a

Write down the first four terms of the sequence defined by

(a) $u_r = \dfrac{r}{r+1}$

(b) $u_1 = 2, u_{r+1} = \dfrac{u_r}{u_r + 1}$

(a) $u_r = \dfrac{r}{r+1} \quad \Rightarrow \quad u_1 = \dfrac{1}{1+1} = \dfrac{1}{2}$

$$u_2 = \dfrac{2}{2+1} = \dfrac{2}{3}$$

$$u_3 = \dfrac{3}{3+1} = \dfrac{3}{4}$$

$$u_4 = \dfrac{4}{4+1} = \dfrac{4}{5}$$

(b) $u_1 = 2$, and $u_{r+1} = \dfrac{u_r}{u_r + 1} \quad \Rightarrow \quad u_2 = \dfrac{2}{2+1} = \dfrac{2}{3}$

$$u_3 = \dfrac{2/3}{2/3 + 1} = \dfrac{2}{5}$$

$$u_4 = \dfrac{2/5}{2/5 + 1} = \dfrac{2}{7}$$

The Behaviour of u_r as $r \to \infty$

Consider the sequence $1/2, 2/3, 3/4, 4/5, \ldots$ All the terms are less than 1, and the values of the terms are increasing as r increases, i.e. as the sequence progresses, the value of the terms is getting closer to 1. Expressing this in symbols we have

$$u_r \to 1 \quad \text{as} \quad r \to \infty \quad \text{or} \quad \lim_{r \to \infty} u_r = 1$$

and we say that the sequence *converges*.

We can illustrate this on a graph by plotting values of u_r against values of r.

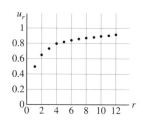

Any sequence whose terms approach one finite value is said to be convergent.

A sequence that is not convergent is called divergent.

Some examples of divergent sequences are

$2, 4, 6, 8, 10, \ldots$ [1]

$1, -1, 1, -1, 1, -1, \ldots$ [2]

The terms in [1] are increasing without limit, so this sequence is clearly divergent.

The terms in [2] form a repeating pattern, i.e. $1, -1$, so they do not approach *one* finite value and the sequence is divergent.

A divergent sequence whose terms form a repeating pattern is called a periodic sequence.

For example, $1, 2, 3, 1, 2, 3, 1, 2, 3, \ldots$ is a periodic sequence. As the sequence progresses, we can describe the behaviour of the terms as periodic, or say that it cycles or oscillates in a regular pattern.

Now consider the sequences $1, -2, 3, -4, 5, -6, \ldots$ [3]

$1, 1/2, 3, 1/4, 5, 1/6, \ldots$ [4]

$2, 3, 1, 1, 2, 2, 1, 3, 1, \ldots$ [5]

Each of these sequences is divergent but none is periodic, nor do the terms in any one of them consistently increase or decrease; they can be said to *oscillate* but not in a repeating pattern. This can be illustrated clearly by plotting values of u_r against values of r. (The broken line has no significance, it is there so that we can see the irregular wave form produced by the progression of the terms.)

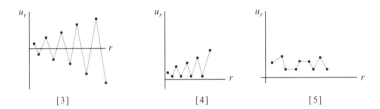

[3] [4] [5]

It is usually obvious from a few terms of a sequence what the nature of that sequence is. If this is not the case, a sketch showing values of u_r plotted against r will usually make it clear. The examples below show such plots for convergent, periodic and simple divergent sequences.

convergent   convergent   periodic   divergent

EXERCISE 18a

1. Write down the first six terms of each sequence and state whether it is convergent, divergent, divergent and periodic, or divergent and oscillating.

(a) $u_r = \dfrac{1}{r^r}$

(b) $u_r = (-1)^r 2^r$

(c) $u_r = \dfrac{(-1)^r}{2^r}$

(d) $u_r = \sin \dfrac{r\pi}{6}$

2. Write down the first six terms of the sequence given by the recurrence relation and determine the behaviour of the sequence as it progresses.

   (a) $u_{r+1} = \dfrac{1 + u_r}{1 + 2u_r}$; $u_1 = 3$

   (b) $u_{r+1} = 2 - (u_r)^2 - u_r$; $u_1 = -4$

   (c) $u_{r+1} = \sqrt{2 - u_r}$; $u_1 = 0.5$

   (d) $u_{r+1} = \frac{1}{5}(5 - \{u_r\}^2)$; $u_1 = 1$

3. Find the nature of the sequence generated by the iteration formula

   $$u_{r+1} = \dfrac{2}{2 - u_r} \quad \text{when}$$

   (a) $u_1 = 1$ (b) $u_1 = 0.5$ (c) $u_1 = -2$

4. Repeat question 3 for the sequence defined by $u_{r+1} = \frac{1}{4}(2 - u_r)$ when

   (a) $u_1 = 1$ (b) $u_1 = -1$ (c) $u_1 = 2$

5. A sequence is generated by the recurrence relation $u_{r+1} = \dfrac{1}{u_r - 3}$

   Given that $u_2 = u_1$ find the possible values of u_1.

SERIES

When the terms of a sequence are added, a series is formed,
e.g., $1 + 2 + 4 + 8 + 16 + \ldots$ is a series.

If the series stops after a finite number of terms it is called a finite series,
e.g., $1 + 2 + 4 + 8 + 16 + 32 + 64$ is a finite series of seven terms.

If the series continues indefinitely it is called an infinite series,
e.g., $1 + \frac{1}{2} + \frac{1}{4} + \frac{1}{8} + \frac{1}{16} + \frac{1}{32} + \ldots + \frac{1}{1024} + \ldots$ is an infinite series.

Consider again the series $1 + 2 + 4 + 8 + 16 + 32 + 64$

As each term is a power of 2 we can write this series in the form

$$2^0 + 2^1 + 2^2 + 2^3 + 2^4 + 2^5 + 2^6$$

All the terms of this series are of the form 2^r, so 2^r is a general term. We can then define the series as the sum of terms of the form 2^r where r takes all integral values in order from 0 to 6 inclusive.

Using Σ as a symbol for 'the sum of terms such as' we can redefine our series more concisely as $\Sigma 2^r$, r taking all integral values from 0 to 6 inclusive, or, even more briefly,

$$\sum_{r=0}^{6} 2^r$$

Placing the lowest and highest value that r takes below and above the sigma symbol respectively, indicates that r takes all integral values between these extreme values.

Thus $\displaystyle\sum_{r=2}^{10} r^3$ means 'the sum of all terms of the form r^3 where r takes all integral values from 2 to 10 inclusive',

i.e. $\displaystyle\sum_{r=2}^{10} r^3 = 2^3 + 3^3 + 4^3 + 5^3 + 6^3 + 7^3 + 8^3 + 9^3 + 10^3$

Note that a finite series, when written out, should always end with the last term even if intermediate terms are omitted, e.g. $3 + 6 + 9 + \ldots + 99$

The infinite series $\quad 1 + \frac{1}{2} + \frac{1}{4} + \frac{1}{8} + \frac{1}{16} + \ldots$

may also be written in the sigma notation. The continuing dots after the last written term indicate that the series is infinite, i.e. there is *no* last term. Each term of this series is a power of $\frac{1}{2}$ so a general term can be written $\left(\frac{1}{2}\right)^r$. The first term is 1 or $\left(\frac{1}{2}\right)^0$, so the first value that r takes is zero. There is no last term of this series, so there is no upper limit for the value of r.

Therefore $\quad 1 + \frac{1}{2} + \frac{1}{4} + \frac{1}{8} + \frac{1}{16} + \ldots$ may be written as $\displaystyle\sum_{r=0}^{\infty} \left(\frac{1}{2}\right)^r$

Note that when a given series is rewritten in the sigma notation it is as well to check that the first few values of r give the correct first few terms of the series.

Writing a series in the sigma notation, apart from the obvious advantage of brevity, allows us to select a particular term of a series without having to write down all the earlier terms.

For example, in the series $\displaystyle\sum_{r=3}^{10} (2r+5)$,

the first term is the value of $2r+5$ when $r=3$, i.e. $2\times 3+5=11$
the last term is the value of $2r+5$ when $r=10$, i.e. 25
the fourth term is the value of $2r+5$ when r takes its fourth value in order
from $r=3$, i.e. when $r=6$

Thus the fourth term of $\displaystyle\sum_{r=3}^{10} (2r+5)$ is $2\times 6+5=17$

Example 18b

Write the following series in the sigma notation,

(a) $1-x+x^2-x^3+\ldots$ (b) $2-4+8-16+\ldots+128$

(a) A general term of this series is $\pm x^r$, having a positive sign when r is even
and a negative sign when r is odd.
Because $(-1)^r$ is positive when r is even and negative when r is odd, the
general term can be written $(-1)^r x^r$

The first term of this series is 1, or x^0

Hence $1-x+x^2-x^3+\ldots = \displaystyle\sum_{r=0}^{\infty} (-1)^r x^r$

(b) $2-4+8-16+\ldots+128 = 2-(2)^2+(2)^3-(2)^4+\ldots+(2)^7$

So a general term is of the form $\pm 2^r$, being positive when r is odd and
negative when r is even,
i.e. the general term is $(-1)^{r+1}2^r$

Hence $2-4+8-16+\ldots+128 = \displaystyle\sum_{r=1}^{7} (-1)^{r+1}2^r$

EXERCISE 18b

1. Write the following series in the sigma notation:

(a) $1+8+27+64+125$ (d) $1+\frac{1}{3}+\frac{1}{9}+\frac{1}{27}+\ldots$

(b) $2+4+6+8+\ldots+20$ (e) $-4-1+2+5\ldots+17$

(c) $\frac{1}{2}+\frac{1}{3}+\frac{1}{4}+\frac{1}{5}+\ldots+\frac{1}{50}$ (f) $8+4+2+1+\frac{1}{2}+\ldots$

2. Write down the first three terms and, where there is one, the last term of each of the following series:

(a) $\displaystyle\sum_{r=1}^{\infty} \frac{1}{r}$

(b) $\displaystyle\sum_{r=0}^{5} r(r+1)$

(c) $\displaystyle\sum_{r=0}^{20} \frac{r+2}{(r+1)(2r+1)}$

(d) $\displaystyle\sum_{r=0}^{\infty} \frac{1}{(r^2+1)}$

(e) $\displaystyle\sum_{r=-1}^{8} r(r+1)(r+2)$

(f) $\displaystyle\sum_{r=0}^{\infty} a^r(-1)^{r+1}$

3. For the following series, write down the term indicated, and the number of terms in the series.

(a) $\displaystyle\sum_{r=1}^{9} 2^r$, 3rd term

(b) $\displaystyle\sum_{r=-1}^{8} (2r+3)$, 5th term

(c) $\displaystyle\sum_{r=-6}^{-1} \frac{1}{(2r+1)}$, last term

(d) $\displaystyle\sum_{r=0}^{\infty} \frac{1}{(r+1)(r+2)}$, 20th term

(e) $\displaystyle\sum_{r=1}^{\infty} \left(\frac{1}{2}\right)^r$, nth term

(f) $8+4+0-4-8-12\ldots-80$, 15th term

(g) $\frac{1}{16}+\frac{1}{8}+\frac{1}{4}+\frac{1}{2}+\ldots+32$, 7th term

ARITHMETIC PROGRESSION

Consider the sequence 5, 8, 11, 14, 17, ..., 29
Each term of this sequence exceeds the previous term by 3, so the sequence can be written in the form

$$5, \ (5+3), \ (5+2\times3), \ (5+3\times3), \ (5+4\times3),\ldots, \ (5+8\times3)$$

This sequence is an example of an arithmetic progression (AP) which is a sequence where any term differs from the preceding term by a constant, called the *common difference*.

The common difference may be positive or negative. For example, the first six terms of an AP whose first term is 8 and whose common difference is -3, are
8, 5, 2, -1, -4, -7

In general, if an AP has a first term a, and a common difference d, the first four terms are $a, (a+d), (a+2d), (a+3d)$,
and the nth term, u_n, is $a+(n-1)d$

Thus   **an AP with n terms can be written as**

$$a, (a+d), (a+2d), \ldots, [a+(n-1)d]$$

Examples 18c

1. The 8th term of an AP is 11 and the 15th term is 21. Find the common difference, the first term of the series, and the nth term.

If the first term of the series is a and the common difference is d, then the 8th term is $a+7d$,

$\therefore \qquad a+7d = 11$ [1]

and the 15th term is $a+14d$,

$\therefore \qquad a+14d = 21$ [2]

$[2]-[1]$ gives $\qquad 7d = 10 \quad \Rightarrow \quad d = \frac{10}{7}$

and $\qquad\qquad\qquad\qquad a = 1$

so the first term is 1 and the common difference is $\frac{10}{7}$

Hence the nth term is $a+(n-1)d = 1+(n-1)\frac{10}{7} = \frac{1}{7}(10n-3)$

2. The nth term of an AP is $12-4n$. Find the first term and the common difference.

If the nth term is $12-4n$, the first term $(n=1)$ is 8
The second term $(n=2)$ is 4

Therefore the common difference is -4

The Sum of an Arithmetic Progression

Consider the sum of the first ten even numbers, which is an AP.

Writing it first in normal, then in reverse, order we have

$$S = 2 + 4 + 6 + 8 + \ldots + 18 + 20$$

$$S = 20 + 18 + 16 + 14 + \ldots + 4 + 2$$

Adding gives $2S = 22 + 22 + 22 = 22 + \ldots + 22 + 22$

As there are ten terms in this series, we have

$$2S = 10 \times 22 \quad \Rightarrow \quad S = 110$$

This process is known as finding the sum from first principles. Applying it to a general AP gives formulae for the sum, which may be quoted and used.

If S_n is the sum of the first n terms of an AP with last term l,

then $S_n = a + (a + d) + (a + 2d) + \ldots + (l - d) + l$

reversing $S_n = l + (l - d) + (l - 2d) + \ldots + (a + d) + a$

adding $2S_n = (a + l) + (a + l) + (a + l) + \ldots + (a + l) + (a + l)$

as there are n terms we have $2S_n = n(a + l)$

$\Rightarrow \quad S_n = \frac{1}{2}n(a + l)$ **i.e. $S_n = (\text{number of terms}) \times (\text{average term})$**

Also, because the nth term, l, is equal to $a + (n - 1)d$, we have

$$S_n = \frac{1}{2}n[a + a + (n - 1)d]$$

i.e. $$S_n = \frac{1}{2}n[2a + (n - 1)d]$$

Either of these formulae can now be used to find the sum of the first n terms of an AP.

Examples 18c (continued)

3. Find the sum of the following series,

   (a) an AP of eleven terms whose first term is 1 and whose last term is 6

   (b) $\displaystyle\sum_{r=1}^{8}\left(2-\frac{2r}{3}\right)$

   (a) We know the first and last terms, and the number of terms so we use $S_n = \frac{1}{2}n(a+l)$

$\Rightarrow$ $S_{11} = \frac{11}{2}(1+6) = \frac{77}{2}$

   (b) $\displaystyle\sum_{r=1}^{8}\left(2-\frac{2r}{3}\right) = \frac{4}{3}+\frac{2}{3}+0-\frac{2}{3}-\ldots-\frac{10}{3}$

   This is an AP with 8 terms where $a = \frac{4}{3}, \ d = -\frac{2}{3}$

   Using $S_n = \frac{1}{2}n[2a+(n-1)d]$ gives

$$S_8 = 4\left[\frac{8}{3}+7\left(-\frac{2}{3}\right)\right] = -8$$

4. In an AP the sum of the first ten terms is 50 and the 5th term is three times the 2nd term. Find the first term and the sum of the first 20 terms.

If a is the first term and d is the common difference, and there are n terms, using $S_n = \frac{1}{2}n[2a+(n-1)d]$ gives

$$S_{10} = 50 = 5(2a+9d) \tag{1}$$

Now using $u_n = a+(n-1)d$ gives

$$u_5 = a+4d \ \text{ and } \ u_2 = a+d$$

Therefore $a+4d = 3(a+d)$ [2]

From [1] and [2] we get $d = 1$ and $a = \frac{1}{2}$
so the first term is $\frac{1}{2}$ and the sum of the first 20 terms is S_{20} where

$$S_{20} = 10(1+19\times1) = 200$$

5. The sum of the first n terms of a series is given by $S_n = n(n+3)$ Find the fourth term of the series and show that the terms are in arithmetic progression.

If the terms of the series are $a_1, a_2, a_3, \ldots a_n$

then $\quad S_n = a_1 + a_2 + \ldots + a_n = n(n+3)$

So $\quad S_4 = a_1 + a_2 + a_3 + a_4 = 28$

and $\quad S_3 = a_1 + a_2 + a_3 \quad = 18$

Hence the fourth term of the series, a_4, is 10

Now $\quad S_n = a_1 + a_2 + \ldots + a_{n-1} + a_n = n(n+3)$

and $\quad S_{n-1} = a_1 + a_2 + \ldots + a_{n-1} \quad = (n-1)(n+2)$

Hence the nth term of the series, a_n, is given by

$$a_n = n(n+3) - (n-1)(n+2) = 2n+2$$

Replacing n by $n-1$ gives the $(n-1)$th term

i.e. $\quad a_{n-1} = 2(n-1)+2 = 2n$

Then $\quad a_n - a_{n-1} = (2n+2) - 2n = 2$

i.e. there is a common difference of 2 between successive terms, showing that the series is an AP.

EXERCISE 18c

1. Write down the fifth term and the nth term of the following APs.

(a) $\displaystyle\sum_{r=1}^{n}(2r-1)$ (b) $\displaystyle\sum_{r=1}^{n}4(r-1)$ (c) $\displaystyle\sum_{r=0}^{n}(3r+3)$

(d) first term 5, common difference 3

(e) first term 6, common difference -2

(f) first term p, common difference q

(g) first term 10, last term 30, 11 terms

(h) $1, 5, \ldots$ (i) $2, 1\frac{1}{2}, \ldots$ (j) $-4, -1, \ldots$

2. Find the sum of the first ten terms of each of the series given in Question (1).

3. The 9th term of an AP is 8 and the 4th term is 20. Find the first term and the common difference.

4. The 6th term of an AP is twice the 3rd term and the first term is 3. Find the common difference and the 10th term.

5. The nth term of an AP is $\frac{1}{2}(3-n)$. Write down the first three terms and the 20th term.

6. Find the sum, to the number of terms indicated, of each of the following APs.

   (a) $1 + 2\frac{1}{2} + \ldots$, 6 terms   (b) $3 + 5 + \ldots$, 8 terms

   (c) the first twenty odd integers   (d) $a_1 + a_2 + a_3 + \ldots + a_8$ where $a_n = 2n + 1$

   (e) $4 + 6 + 8 + \ldots + 20$ (f) $\displaystyle\sum_{r=1}^{3n} (3 - 4r)$

   (g) $S_n = n^2 - 3n$, 8 terms   (h) $S_n = 2n(n+3)$, m terms

7. The sum of the first n terms of an AP is S_n where $S_n = n^2 - 3n$. Write down the fourth term and the nth term.

8. The sum of the first n terms of a series is given by S_n where $S_n = n(3n - 4)$. Show that the terms of the series are in arithmetic progression.

9. In an arithmetic progression, the 8th term is twice the 4th term and the 20th term is 40. Find the common difference and the sum of the terms from the 8th to the 20th inclusive.

10. How many terms of the AP, $1 + 3 + 5 + \ldots$ are required to make a sum of 1521?

11. Find the least number of terms of the AP, $1 + 3 + 5 + \ldots$ that are required to make a sum exceeding 4000.

12. If the sum of the first n terms of a series is S_n where $S_n = 2n^2 - n$,

    (a) prove that the series is an AP, stating the first term and the common difference,

    (b) find the sum of the terms from the 3rd to the 12th inclusive.

13. In an AP the 6th term is half the 4th term and the 3rd term is 15.

    (a) Find the first term and the common difference.

    (b) How many terms are needed to give a sum that is less than 65?

GEOMETRIC PROGRESSIONS

Consider the sequence

$$12, 6, 3, 1.5, 0.75, 0.375, \ldots$$

Each term of this sequence is half the preceeding term so the sequence may be written

$$12, \ 12\left(\tfrac{1}{2}\right), \ 12\left(\tfrac{1}{2}\right)^2, \ 12\left(\tfrac{1}{2}\right)^3, \ 12\left(\tfrac{1}{2}\right)^4, \ 12\left(\tfrac{1}{2}\right)^5, \ldots$$

Such a sequence is called a geometric progression (GP) which is a sequence where each term is a constant multiple of the preceding term. This constant multiplying factor is called the common ratio, and it may have any real value.

Hence, if a GP has a first term of 3 and a common ratio of -2 the first four terms are

$$3, \ 3(-2), \ 3(-2)^2, \ 3(-2)^3$$

or $\qquad 3, \ -6, \ 12, \ -24$

In general if a GP has a first term a, and a common ratio r, the first four terms are

$$a, \ ar, \ ar^2, \ ar^3$$

and the nth term, u_n, is ar^{n-1}, thus

a GP with n terms can be written $a, ar, ar^2, \ldots, ar^{n-1}$

The Sum of a Geometric Progression

Consider the sum of the first eight terms, S_8, of the GP with first term 1 and common ratio 3

i.e. $\qquad\qquad S_8 = 1 + 1(3) + 1(3)^2 + 1(3)^3 + \ldots + 1(3)^7$

$\Rightarrow \qquad\qquad 3S_8 = \qquad\quad 3 + \quad 3^2 + \quad 3^3 + \ldots + \quad 3^7 + 3^8$

Hence $\qquad S_8 - 3S_8 = 1 + \quad 0 + \quad 0 + \quad 0 + \ldots + \quad 0 - 3^8$

So $\qquad S_8(1-3) = 1 - 3^8$

$\Rightarrow \qquad\qquad S_8 = \dfrac{1 - 3^8}{1 - 3} = \dfrac{3^8 - 1}{2}$

This process can be applied to a general GP.

Consider the sum, S_n, of the first n terms of a GP with first term a and common ratio r,

i.e. $S_n = a + ar + \ldots + ar^{n-2} + ar^{n-1}$

Multiplying by r gives

$$rS_n = ar + ar^2 + \ldots + ar^{n-1} + ar^n$$

Hence $S_n - rS_n = a - ar^n$

$\Rightarrow$ $S_n(1 - r) = a(1 - r^n)$

$\Rightarrow$ $S_n = \dfrac{a(1 - r^n)}{1 - r}$

If $r > 1$ the formula may be written $\dfrac{a(r^n - 1)}{r - 1}$

Examples 18d

1. The 5th term of a GP is 8, the third term is 4, and the sum of the first ten terms is positive. Find the first term, the common ratio, and the sum of the first ten terms.

For a first term a and common ratio r, the nth term is ar^{n-1}

Thus we have $ar^4 = 8$ $(n = 5)$

and $ar^2 = 4$ $(n = 3)$

dividing gives $r^2 = 2$

$\Rightarrow$ $r = \pm\sqrt{2}$ and $a = 2$

Using the formula $S_n = \dfrac{a(r^n - 1)}{r - 1}$ gives,

when $r = \sqrt{2}$, $S_{10} = \dfrac{2[(\sqrt{2})^{10} - 1]}{\sqrt{2} - 1} = \dfrac{62}{\sqrt{2} - 1}$

when $r = -\sqrt{2}$, $S_{10} = \dfrac{2[(-\sqrt{2})^{10} - 1]}{-\sqrt{2} - 1} = \dfrac{-62}{\sqrt{2} + 1}$

But we are told that $S_{10} > 0$, so we deduce that

$r = \sqrt{2}$ and $S_{10} = \dfrac{62}{\sqrt{2} - 1} = 62(\sqrt{2} + 1)$

2. A "Save As You Earn" scheme involves paying in £50 on the first day of each month. Interest of 1% of the total in the scheme is added at the end of each month. If the first payment is made on January 1st, find the total amount in the scheme on December 31st of the same year.

If P_1 is the total amount at the end of January, P_2 is the total amount at the end of February, and so on,

then $P_1 = £\{50 + 50(0.01)\} = £(50)(1.01)$

$P_2 = £(50 + P_1)(1.01) = £\{(50)(1.01) + (50)(1.01)^2\}$

$P_3 = £(50 + P_2)(1.01) = £\{(50)(1.01) + (50)(1.01)^2 + (50)(1.01)^3\}$

and so on, giving

$P_{12} = £\{(50)(1.01) + (50)(1.01)^2 + \ldots + (50)(1.01)^{12}\}$

$= £(50)(1.01)[1 + (1.01) + (1.01)^2 + \ldots + (1.01)^{11}]$

The expression in square brackets is a GP of 12 terms with $a = 1$ and $r = 1.01$, hence

$$P_{12} = £(50)(1.01)\left[\frac{(1.01)^{12} - 1}{1.01 - 1}\right]$$

$= £640.47$ to the nearest penny.

3. The sum of the first n terms of a series is 3^{n-1}. Show that the terms of this series are in geometric progression and find the first term, the common ratio and the sum of the second n terms of this series.

If the series is $\quad a_1 + a_2 + \ldots + a_n$

then $\quad S_n = a_1 + a_2 + \ldots + a_{n-1} + a_n = 3^n - 1$

and $\quad S_{n-1} = a_1 + a_2 + \ldots + a_{n-1} \quad = 3^{n-1} - 1$

therefore $\quad a_n = 3^n - 1 - (3^{n-1} - 1)$

i.e. the nth term is $3^n - 3^{n-1} = 3^{n-1}(3 - 1) = (2)3^{n-1}$

Similarly $\quad a_{n-1} = (2)3^{n-2}$ so $a_n \div a_{n-1} = 3$

showing that successive terms in the series have a constant ratio of 3. Hence this series is a GP with first term 2 and common ratio 3.

The sum of the second n terms is

(the sum of the first $2n$ terms) − (the sum of the first n terms)

$= S_{2n} - S_n$

$= (3^{2n} - 1) - (3^n - 1)$

$= 3^n(3^n - 1)$

EXERCISE 18d

1. Write down the fifth term and the nth term of the following GPs:

   (a) $2, 4, 8, \dots$ (b) $2, 1, \frac{1}{2} \dots$ (c) $3, -6, 12, \dots$

   (d)   first term 8, common ratio $-\frac{1}{2}$

   (e)   first term 3, last term $\frac{1}{81}$, 6 terms

2. Find the sum, to the number of terms given, of the following GPs.

   (a) $3 + 6 + \dots$, 6 terms (b) $3 - 6 + \dots$, 8 terms

   (c) $1 + \frac{1}{2} + \frac{1}{4} + \dots$, 20 terms

   (d)   first term 5, common ratio $\frac{1}{5}$, 5 terms

   (e)   first term $\frac{1}{2}$, common ratio $-\frac{1}{2}$, 10 terms

   (f)   first term 1, common ratio -1, 2001 terms.

3. The 6th term of a GP is 16 and the 3rd term is 2. Find the first term and the common ratio.

4. Find the common ratio, given that it is negative, of a GP whose first term is 8 and whose 5th term is $\frac{1}{2}$.

5. The nth term of a GP is $\left(-\frac{1}{2}\right)^n$. Write down the first term and the 10th term.

6. Evaluate $\displaystyle\sum_{r=1}^{10} (1.05)^r$

7. Find the sum to n terms of the following series.

   (a) $x + x^2 + x^3 + \dots$ (b) $x + 1 + \dfrac{1}{x} + \dots$ (c) $1 - y + y^2 - \dots$

   (d) $x + \dfrac{x^2}{2} + \dfrac{x^3}{4} + \dfrac{x^4}{8} + \dots$ (e) $1 - 2x + 4x^2 - 8x^3 + \dots$

8. Find the sum of the first n terms of the GP $2 + \frac{1}{2} + \frac{1}{8} + \dots$ and find the least value of n for which this sum exceeds 2.65

9. The sum of the first 3 terms of a GP is 14. If the first term is 2, find the possible values of the sum of the first 5 terms.

10. Evaluate $\displaystyle\sum_{r=1}^{10} 3(3/4)^r$

1. A mortgage is taken out for £10 000 and is repaid by annual instalments of £2000. Interest is charged on the outstanding debt at 10%, calculated annually. If the first repayment is made one year after the mortgage is taken out find the number of years it takes for the mortgage to be repaid.

2. A bank loan of £500 is arranged to be repaid in two years by equal monthly instalments. Interest, *calculated monthly*, is charged at 11% p.a. on the remaining debt. Calculate the monthly repayment if the first repayment is to be made one month after the loan is granted.

CONVERGENCE OF SERIES

If a piece of string, of length l, is cut up by first cutting it in half and keeping one piece, then cutting the remainder in half and keeping one piece, then cutting the remainder in half and keeping one piece, and so on, the sum of the lengths retained is

$$\frac{l}{2} + \frac{l}{4} + \frac{l}{8} + \frac{l}{16} + \ldots$$

As this process can (in theory) be carried on indefinitely, the series formed above is infinite.

After several cuts have been made the remaining part of the string will be very small indeed, so the sum of the cut lengths will be very nearly equal to the total length, l, of the original piece of string. The more cuts that are made the closer to l this sum becomes,

i.e. if after n cuts, the sum of the cut lengths is

$$\frac{l}{2} + \frac{l}{2^2} + \frac{l}{2^3} + \ldots + \frac{l}{2^n}$$

then, as $n \to \infty$, $\quad \dfrac{l}{2} + \dfrac{l}{2^2} + \ldots + \dfrac{l}{2^n} \to l$

or $\quad \lim_{n \to \infty} \left[\dfrac{l}{2} + \dfrac{l}{2^2} + \ldots + \dfrac{l}{2^n} \right] = l$

l is called the sum to infinity of this series.

In general, if S_n is the sum of the first n terms of any series and if $\lim\limits_{n \to \infty} [S_n]$ exists and is finite, the series is said to be *convergent*.

In this case the sum to infinity, S_∞, is given by

$$S_\infty = \lim_{n \to \infty} [S_n]$$

The series $1/2 + 1/2^2 + 1/2^3 + \ldots$, for example, is convergent as its sum to infinity is l.

However, for the series $1 + 2 + 3 + \ldots + n$, we have $S_n = \frac{1}{2}n(n+1)$ As $n \to \infty$, $S_n \to \infty$ so this series does not converge and is said to be divergent.

For any AP, $S_n = \frac{1}{2}n[2a + (n-1)d]$, which always approaches infinity as $n \to \infty$. Therefore any AP is divergent.

THE SUM TO INFINITY OF A GP

Consider the general GP $a + ar + ar^2 + \ldots$

Now $S_n = \dfrac{a(1 - r^n)}{1 - r}$, and if $|r| < 1$, then $\lim\limits_{n \to \infty} r^n = 0$

So $\lim\limits_{n \to \infty} S_n = \lim\limits_{n \to \infty} \left[\dfrac{a(1 - r^n)}{1 - r} \right] = \dfrac{a}{1 - r}$

If $|r| \geqslant 1$, the series does not converge.

Therefore, provided that $|r| < 1$, a GP converges to a sum of $\dfrac{a}{1 - r}$

i.e. **the sum to infinity of a GP is** $\dfrac{a}{1 - r}$

provided that $|r| < 1$

Arithmetic Mean

If three numbers, p_1, p_2, p_3, are in arithmetic progression then p_2 is called the *arithmetic mean* of p_1 and p_3

If $p_1 = a$, we may write p_2, p_3 as $a + d$, $a + 2d$ respectively,

hence $p_1 + p_3 = 2a + 2d = 2(a + d) = 2p_2 \Rightarrow p_2 = \frac{1}{2}(p_1 + p_3)$

i.e. **the arithmetic mean of two numbers m and n is** $\frac{1}{2}(m + n)$

Examples 18e

1. Determine whether each series converges. If it does, give its sum to infinity.

(a) $3 + 5 + 7 + \ldots$ (b) $1 - \frac{1}{4} + \frac{1}{16} - \frac{1}{64} + \ldots$ (c) $3 + \frac{9}{2} + \frac{27}{4} + \ldots$

(a) $3 + 5 + 7 + \ldots$ is an AP $(d = 2)$ and so does not converge.

(b) $1 - \frac{1}{4} + \frac{1}{16} - \frac{1}{64} + \ldots = 1 + (-\frac{1}{4}) + (-\frac{1}{4})^2 + (-\frac{1}{4})^3 + \ldots$

which is a CP where $r = -\frac{1}{4}$, i.e. $|r| < 1$

So this series converges and $S_\infty = \dfrac{a}{1 - r} = \dfrac{1}{1 - (-\frac{1}{4})} = \dfrac{4}{5}$

(c) $3 + \frac{9}{2} + \frac{27}{4} + \ldots = 3 + 3(\frac{3}{2}) + 3(\frac{9}{4}) + \ldots = 3 + 3(\frac{3}{2}) + 3(\frac{3}{2})^2 + \ldots$

This series is a GP where $r = \frac{3}{2}$ and, as $|r| > 1$, the series does not converge.

2. Express the recurring decimal $0.1\dot{5}7\dot{6}$ as a fraction in its lowest terms.

$$0.1\dot{5}7\dot{6} = 0.1\,576\,576\,576\,576\ldots$$

$$= 0.1 + 0.0576 + 0.000\,0576 + 0.000\,000\,0576 + \ldots$$

$$= \frac{1}{10} + \frac{576}{10^4} + \frac{576}{10^7} + \frac{576}{10^{10}} + \ldots$$

$$= \frac{1}{10} + \frac{576}{10^4}\left[1 + \frac{1}{10^3} + \frac{1}{10^6} + \ldots\right]$$

$$= \frac{1}{10} + \frac{576}{10^4}\left[1 + \frac{1}{10^3} + \left(\frac{1}{10^3}\right)^2 + \ldots\right]$$

Now the series in the square bracket is a GP whose first term is 1, and whose common ratio is $\dfrac{1}{10^3}$.

Hence it has a sum to infinity of $\dfrac{1}{1 - 10^{-3}} = \dfrac{10^3}{999}$

$\Rightarrow \quad 0.1\dot{5}7\dot{6} = \dfrac{1}{10} + \dfrac{576}{10^4} \times \dfrac{10^3}{999} = \dfrac{1}{10} + \dfrac{576}{9990} = \dfrac{1575}{9990} = \dfrac{35}{222}$

3. The 3rd term of a convergent GP is the arithmetic mean of the 1st and 2nd terms.
Find the common ratio and, if the first term is 1, find, the sum to infinity.

If the series is $a + ar + ar^2 + ar^3 + \ldots$

then
$$ar^2 = \tfrac{1}{2}(a + ar)$$

$a \neq 0$, so
$$2r^2 - r - 1 = 0$$

$\Rightarrow$
$$(2r + 1)(r - 1) = 0$$

i.e.
$$r = -\tfrac{1}{2} \text{ or } 1$$

As the series is convergent, the common ratio is $-\tfrac{1}{2}$

When $r = -\tfrac{1}{2}$ and $a = 1$,

$$S_\infty = \frac{1}{1 + \tfrac{1}{2}} = \tfrac{2}{3}$$

EXERCISE 18e

1. Determine whether each of the series given below converges.

(a) $4 + \dfrac{4}{3} + \dfrac{4}{3^2} + \ldots$

(b) $9 + 7 + 5 + 3 + \ldots$

(c) $20 - 10 + 5 - 2.5 + \ldots$

(d) $\dfrac{5}{10} + \dfrac{5}{100} + \dfrac{5}{1000} + \ldots$

(e) $p + 2p + 3p + \ldots$

(f) $3 - 1 + \dfrac{1}{3} - \dfrac{1}{9} + \ldots$

2. Find the sum to infinity of those series in Question 1 that are convergent.

3. Express the following recurring decimals as fractions

(a) $0.16\dot{2}$ (b) $0.\dot{3}\dot{4}$ (c) $0.0\dot{2}\dot{1}$

4. The sum to infinity of a GP is twice the first term. Find the common ratio.

5. The sum to infinity of a GP is 16 and the sum of the first 4 terms is 15. Find the first four terms.

6. If a, b and c are the first three terms of a GP, prove that $\sqrt{a}$, $\sqrt{b}$ and $\sqrt{c}$ form another GP.

MIXED EXERCISE 18

In questions 1 to 9, write down the first five terms of the sequence and describe the behaviour of the terms as the sequence progresses.

1. $u_r = \dfrac{1}{r^2}$

2. $u_r = \dfrac{r-2}{r+2}$

3. $u_r = \cos r\pi$

4. $u_r = (-1)^r$

5. $u_{r+1} = u_r + 2, \quad u_1 = 1$

6. $u_{r+1} = 2u_r, \quad u_1 = 1$

7. $u_{r+1} = (u_r)^{1/2}, \quad u_1 = 4$

8. $u_{r+2} = u_{r+1} + u_r, \quad u_1 = 1, \; u_2 = 2$

9. $u_{r+2} = \dfrac{u_{r+1}}{u_r}, \quad u_1 = 1, \; u_2 = 2$

In Questions 10 to 18 find the sum of each series.

10. $1 - \frac{1}{2} + \frac{1}{4} - \frac{1}{8} + \ldots$

11. $2 - (2)(3) + (2)(3)^2 - (2)(3)^3 + \ldots + (2)(3)^{10}$

12. $\displaystyle\sum_{r=2}^{n} ab^{2r}$

13. $\displaystyle\sum_{r=5}^{n} 4r$

14. $e + e^2 + e^3 + \ldots + e^n$

15. $\displaystyle\sum_{r=1}^{n} (2 + 3r)$

16. $\displaystyle\sum_{r=n}^{2n} (1 - 2r)$

17. $\displaystyle\sum_{r=1}^{\infty} \frac{1}{2^r}$

18. The sum of the first n even numbers.

In Questions 19 and 20, express each decimal as a fraction in its lowest terms.

19. $0.05\dot{1}$ **20.** $0.\dot{1}\dot{0}$

21. The sum of the first n terms of a series is n^3. Write down the first four terms and the nth term of the series.

22. The fourth term of an AP is 8 and the sum of the first ten terms is 40. Find the first term and the tenth term.

23. The second, fourth, and eighth terms of an AP are the first three terms of a GP. Find the common ratio of the GP.

24. Find the value of x for which the numbers $x + 1$, $x + 3$, $x + 7$, are in geometric progression.

25. The second term of a GP is $\frac{1}{2}$ and the sum to infinity of the series is 4. Find the first term and the common ratio of the series.

26. £2000 is invested in a pension fund on January 1st each year. Interest at 9% p.a. is added to the fund on December 31st each year. Calculate the amount in the fund on December 31st 2020 if the first payment was made on January 1st 1980.

27. Jane Smith takes out an endowment policy which involves making a fixed payment each year for 10 years. At the end of the 10 years Jane receives a sum of money equal to her total payments together with interest which is added at the rate of 8% p.a. of the total sum in the fund. Jane will get a payout of £100 000. What is her annual payment?

28. The cost of erecting a tower is £10 000 for the first 10 metres, £25 000 for the next 10 metres, £40 000 for the next 10 metres, £55 000 for the next 10 metres, and so on. Find the cost of erecting a tower 100 metres high.

*29. The population of a new town is projected to grow at the rate of 2.5% a year. How many years is it expected to take for the population to double?

CHAPTER 19

EXPONENTIAL AND LOGARITHMIC FUNCTIONS

EXPONENTIAL FUNCTIONS

Exponent is another word for index or power.

An exponential function is one where the variable is in the index.

For example, 2^x, 3^{-x}, 10^{x+1} are exponential functions of x

Consider the function $f(x) = 2^x$ for which a table of corresponding values of x and $f(x)$ and a graph are given below.

x	$-\infty \leftarrow \ldots$	-10	-1	$-\frac{1}{10}$	0	$\frac{1}{10}$	1	10	$\ldots \rightarrow \infty$
$f(x)$	$0 \leftarrow \ldots$	$\frac{1}{1024}$	$\frac{1}{2}$	0.93	1	1.07	2	1024	$\ldots \rightarrow \infty$

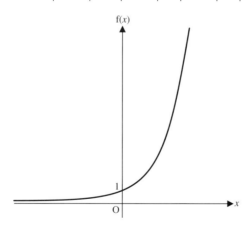

From these we see that
- 2^x has a real value for all real values of x and 2^x is positive for all values of x, i.e. the range of f is $f(x) > 0$
- As $x \to -\infty$, $f(x) \to 0$, i.e. the x-axis is an asymptote.
- As x increases, $f(x)$ increases at a rapidly accelerating rate.

Note also that the curve crosses the y-axis at $(0, 1)$, i.e. $f(0) = 1$. In fact for any function of the form $f(x) = a^x$, where a is a constant and greater than 1, $f(0) = 1$, and the curve representing it is similar in shape to that for 2^x.

EXERCISE 19a

1. On the same set of axes *sketch* the graphs of

   (a) 2^x

   (b) 2^{2x}

   (c) 2^{4x}

2. Draw sketch graphs of the functions

   (a) 3^x

   (b) 4^x

3. (a) Compare your sketches for 1(b) and 2(b) and comment on them.

   (b) Compare your sketches for 1(a), 2(a) and (b). Comment on how the value of a (> 1)affects the shape of the graph of $y = a^x$

4. Write down the values of $f(x) = (1/2)^x$ when $x = -4, -3, -2, -1, 0, 1, 2,$ 3 and 4. From these values deduce the behaviour of $f(x)$ as $x \to \pm\infty$ and hence sketch the graph of the function.

5. By writing $(1/2)^x$ as 2^{-x} explain how the shape of the curve $y = (1/2)^x$ can be deduced from the shape of the curve $y = 2^x$.

6. If $f(x) = 10^x$, sketch the following curves on the same set of axes.

   (a) $y = f(x)$

   (b) $y = f(x+3)$

   (c) $y = f^{-1}(x)$

7. The function f is given by $f: x \to 2^{(3x-2)}$.

   (a) Find $g(x)$ and $h(x)$ such that $f = gh$.

   (b) Evaluate $ff(2)$

   (c) Explain how a sketch of the graph of $y = f(x)$ can be obtained by considering transformations of the curve $y = 2^x$.

EXPONENTIAL GROWTH AND DECAY

There are many situations where a quantity grows or decays by a constant factor over equal time intervals.

For example, if a debt of £100 has 2% interest added each month then, if no repayment is made,

the debt grows to £100 × 1.02 after one month,

$$\text{to } £100 \times 1.02 \times 1.02, \text{ i.e. } £100 \times (1.02)^2 \text{ after two months,}$$

and so on,   to $£100 \times (1.02)^n$ after n months.

1.02 is called the *growth factor* per month.

So if £y is the debt after x months, $y = 100(1.02)^x$.

This is an exponential function that *increases* in value as the exponent, x, increases; we say that y grows exponentially.

Some business assets (e.g. vehicles) depreciate over time. If the value of a lorry is 'written off' by half its value at the start of the year each year, then starting with a value of £A, it is worth £$A \times 0.5$ after one year, £$A \times (0.5)^2$ after two years, and so on.

If £y is the value after x years, then $y = A \times (0.5)^x = A(\frac{1}{2})^x = A \times 2^{-x}$.

Now 2^{-x} is an exponential function that *decreases* in value as x increases, so we say that y decays exponentially.

In each case considered above, the mathematical expression for the relationship is obtained by making certain assumptions and is not necessarily valid at all times. The first example assumes that interest rates remain constant, and we know this is not the case. So the relationship is valid only for the time during which the interest rate is 2%. The assumption in the second case is that the rule for writing down the value of assets never changes and again this is not true in practice; after a few years the vehicle will have a value small enough to be written off completely even if it is not lost through other causes in the mean time.

There are many other situations which, while not giving precisely exact relationships like those above, approximate very closely to exponential growth and decay. So exponential functions are important because they provide mathematical models for situations like these.

THE EXPONENTIAL FUNCTION

The general shape of an exponential curve was seen in the last section. The next diagram shows a few more members of the exponential family.

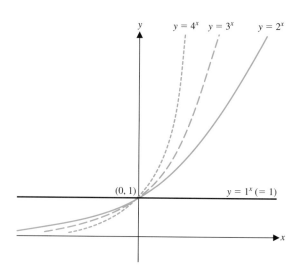

Note that these curves and, in fact, *all* exponential curves, pass through the point $(0, 1)$.

This is because, for any positive base a,

when x is 0, $y = a^x = a^0 = 1$

Each exponential curve has a unique property which the reader can discover experimentally by using an accurate plot of the curve $y = 2^x$. Choose three or four points on the curve and, at each one, draw the tangent as accurately as possible and determine its gradient. Then complete the following table.

Point	Gradient of tangent, i.e. $\dfrac{dy}{dx}$	y-coordinate	$\dfrac{dy}{dx} \div y$
1			
2			
3			
4			

An accurate drawing should result in numbers in the last column that are all reasonably close to 0.7

When this experiment is carried out for 3^x and 4^x we find again that $\dfrac{dy}{dx} \div y$ has a constant value; for 3^x the constant is about 1.1 and for 4^x it is about 1.4
So we have

Base	2	3	4
$\dfrac{dy}{dx} \div y$	0.7	1.1	1.4

⇒

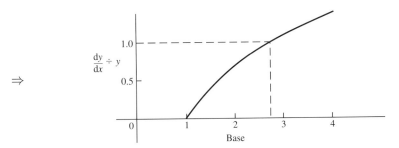

From this graph it can be seen that there is a base, somewhere between 2 and 3, for which $\dfrac{dy}{dx} \div y = 1$, i.e. $\dfrac{dy}{dx} = y$

By calling this base e we see that

$$\text{if } y = e^x \text{ then } \frac{dy}{dx} = e^x$$

The function e^x is the only function which is unchanged when differentiated.

In the early eighteenth century, a number of mathematicians, working along different lines of investigation, all discovered the number e at about the same time.

The number e is irrational, i.e. like π, $\sqrt{2}$, etc., it cannot be given an exact decimal value but, to 4 significant figures, e = 2.718   The value of various powers of e, such as e^2, e^3, e^4, can be obtained from a calculator.

Summing up:

for any value of $a\,(a > 0)$, a^x is *an* exponential function

for the base e ($e \simeq 2.718$), e^x is *the* exponential function

$$\frac{d}{dx}(e^x) = e^x$$

The following diagrams show sketches of $y = e^x$ and of some simple variations.

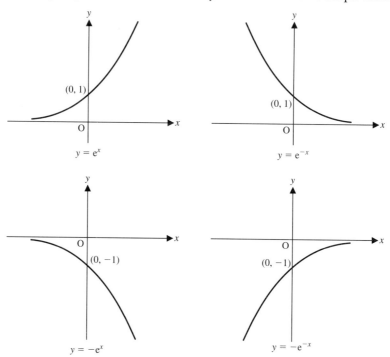

Example 19b

Find the coordinates of the stationary point on the curve $y = e^x - x$, and determine its type. Sketch the curve showing the stationary point clearly.

$$y = e^x - x \quad \Rightarrow \quad \frac{dy}{dx} = e^x - 1$$

At a stationary point, $\dfrac{dy}{dx} = 0$ therefore $e^x - 1 = 0$

i.e. $\quad e^x = 1 \quad \Rightarrow \quad x = 0$

When $x = 0$, $y = e^0 - 0 = 1$

Therefore $(0, 1)$ is a stationary point.

x	-1	0	1
$\dfrac{dy}{dx}$	$-$ve	0	$+$ve

Therefore $(0, 1)$ is a minimum point.

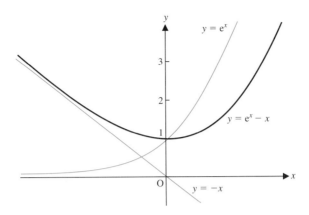

This curve is deduced from separate sketches of $y = e^x$ and $y = -x$ by adding their ordinates.

EXERCISE 19b

1. Evaluate, correct to 3 sf

   (a) e^2 (b) e^{-1} (c) $e^{1.5}$ (d) $e^{-0.3}$

2. Write down the derivative of

   (a) $2e^x$ (b) $x^2 - e^x$ (c) e^x

   In questions 3 to 5 find the gradient of each curve at the specified value of x

3. $y = e^x - 2x$ where $x = 2$

4. $y = x^2 + 2e^x$ where $x = 1$

5. $y = e^x - 3x^3$ where $x = 0$

6. Find the value of x at which the function $e^x - 2x$ has a stationary value.

7. Sketch each given curve.

   (a) $y = 1 - e^x$ (b) $y = e^x + 1$

   (c) $y = 1 - e^{-x}$ (d) $y = 1 + e^{-x}$

8. A culture dish was seeded with $4\,mm^2$ of mould. The table shows the area of the mould at six-hourly intervals.

Time, x hrs	0	6	12	18	24	30	36
Area, $y\,mm^2$	4	8.1	15.9	33	68	118	190

(a) Show that, for a time, the growth factor in the area for each interval of 6 hours is roughly 2 (i.e. the area approximately doubles every 6 hours). Hence show that it is reasonable to use an exponential function as a model for the area and suggest a possible function.

(b) On the same set of axes, plot the points in the table and draw the curve given by the equation you chose as the model.

(c) Give reasons why (i) the model is approximate (ii) when and why the model ceases to be reasonable.

THE LOGARITHMIC FUNCTION

If we consider the function $f : x \to e^x$, then the inverse mapping is given by $e^x \to x$.

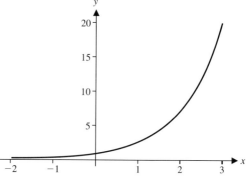

From the graph of $f(x) = e^x$, we can see that one value of e^x maps to one value of x, so the mapping $e^x \to x$ is a function,

i.e., if $f(x) = e^x$

then f^{-1} exists.

Now the equation of the function f is $y = e^x$ and the curve $y = f^{-1}(x)$ is obtained by reflecting the curve $y = e^x$ in the line $y = x$.

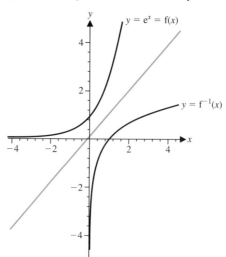

The equation of the reflected curve is $e^y = x$,
i.e. y is the power to which e is raised to give x.
Using 'ln' to mean 'the power to which e is raised to give',
we have $\qquad y = \ln x$

Therefore $\qquad$ if $\qquad$ $f : x \rightarrow e^x$

$\qquad\qquad\qquad$ then $\quad f^{-1} : x \rightarrow \ln x$

The function ln is called *the logarithmic function* and $\ln x$ is called *the natural logarithm* of x.

From the graph of $y = \ln x$ we see that $\ln x$ exists only for positive values of x, so

$$\textbf{ln } x \textbf{ is undefined for negative values of } x.$$

Hence the domain of the logarithmic function is $x > 0$,
and values of $\ln x$ range from $-\infty$ to ∞.

Particular values of $\ln x$ can be found using a scientific calculator.

The Derivative of ln x

We know that $\quad y = e^x \quad \Leftrightarrow \quad x = e^y \quad$ and we also know how to differentiate the exponential function.

So a relationship between $\dfrac{d}{dx}(y)$ and $\dfrac{d}{dy}(x)$ would help in finding the derivative of $\ln x$.

Consider the equation $y = f(x)$ where $f(x)$ is any function of x.

$$\frac{dy}{dx} = \lim_{\delta x \to 0} \frac{\delta y}{\delta x} = \lim_{\delta x \to 0} 1 \Big/ \left(\frac{\delta x}{\delta y} \right)$$

Now $\delta y \to 0$ as $\delta x \to 0$

$$\therefore \qquad \frac{dy}{dx} = \lim_{\delta x \to 0} 1 \Big/ \left(\frac{\delta x}{\delta y} \right)$$

i.e. $\qquad \dfrac{dy}{dx} = 1 \Big/ \left(\dfrac{dx}{dy} \right)$

This relationship can be used to find the derivative of *any* function if the derivative of the inverse is known. We will now apply it to differentiate $\ln x$.

$$y = \ln x \quad \Leftrightarrow \quad x = e^y$$

Differentiating e^y w.r.t. to y gives

$$\frac{dx}{dy} = e^y = x$$

Therefore $\quad \dfrac{dy}{dx} = 1 / \dfrac{dx}{dy} = \dfrac{1}{x}$

i.e. $\qquad\qquad\qquad\qquad \dfrac{d}{dx} \ln x = \dfrac{1}{x}$

We can use this result to find the derivative of e^{ax} where a is a constant.

If $\quad y = e^{ax}\quad$ then $\quad ax = \ln y \quad \Rightarrow \quad x = \dfrac{1}{a}\ln y$

$\therefore \qquad \dfrac{dx}{dy} = \dfrac{1}{a} \times \dfrac{1}{y} = \dfrac{1}{ay}$

Using $\dfrac{dy}{dx} = 1/\dfrac{dx}{dy}$ gives $\dfrac{dy}{dx} = ay = a e^{ax}$

i.e. $\qquad\qquad\qquad\qquad \dfrac{d}{dx}(e^{ax}) = a e^{ax}$

Example 19c

Find the derivative of   (a) $3 \ln x$ (b) $1/e^{4x}$

(a) If $y = 3 \ln x$, $\qquad \dfrac{dy}{dx} = 3 \times \dfrac{1}{x} = \dfrac{3}{x}$

(b) $y = 1/e^{4x} \quad \Rightarrow \quad y = e^{-4x}$, $\qquad \dfrac{dy}{dx} = -4e^{-4x}$

EXERCISE 19c

Find, correct to 3 significant figures, the value of

1. $\ln 2$ 2. $\ln 1.15$ 3. $\ln 0.5$ 4. $\ln 17.3$

In questions 5 and 6, find the gradient of the curve at the given value of x.

5. (a) $y = \ln x, \quad x = 1$ (b) $y = 1 + \ln x, \quad x = 0.4$
    (c) $y = x - \ln x, \quad x = 1/2$ (d) $y = x + 3 \ln x, \quad x = 1.5$

6. (a) $y = e^{3x}$, $x = 0$ (b) $y = x + e^{2x}$, $x = 1$

   (c) $y = e^{-4x}$, $x = 0.5$ (d) $y = x - e^{-x}$, $x = 2$

7. Find the derivative of

   (a) $\dfrac{1}{e^{3x}}$ (b) $\dfrac{1 - e^x}{e^x}$

Find the coordinates of the stationary points on each curve.

8. $y = x - \ln x$ 9. $y = \ln x - \sqrt{x}$ 10. $y = e^x - x$

Sketch each of the following curves.

11. $y = -\ln x$ 12. $y = \ln(-x)$ 13. $y = 2 + \ln x$

Integration

We know that $\dfrac{d}{dx}(e^{ax}) = ae^{ax}$, therefore $\int ae^{ax}\,dx = e^{ax}$

$\Rightarrow$ $\int e^{ax}\,dx = \dfrac{1}{a}e^{ax} + k$

Similarly $\dfrac{d}{dx}2\ln x = \dfrac{2}{x}$ so $\int \dfrac{2}{x}\,dx = 2\ln x + k$

$\therefore$ $\int \dfrac{a}{x}\,dx = a\ln x + k$

Examples 19d

1. Evaluate $\displaystyle\int_1^2 \left(\dfrac{5}{x}\right)dx$

$\displaystyle\int_1^2 \left(\dfrac{5}{x}\right)dx = \Big[\,5\ln x\,\Big]_1^2$

$= (5\ln 2) - (5(0)) = 5\ln 2$

The answer is left in exact form; if an answer is required correct to 3 sf, say, this expression can be evaluated using a calculator. There are calculators on the market that give numerical values of definite integrals, however these are not suitable for use if an exact answer is required.

2. Evaluate $\displaystyle\int_1^2 2e^{4x}\,dx$

$$\int_1^2 2e^{4x}dx = \left[2(\tfrac14 e^{4x})\right]_1^2$$
$$= \tfrac12 e^8 - \tfrac12 e^4 = \tfrac12 e^4(e^4 - 1)$$

EXERCISE 19d

Find

1. $\displaystyle\int 2e^x dx$

2. $\displaystyle\int e^{-x}dx$

3. $\displaystyle\int e^{3x}dx$

4. $\displaystyle\int (2 - e^x)\,dx$

5. $\displaystyle\int (e^x + 1)^2 dx$

6. $\displaystyle\int e^{-x}(1 - e^{2x})\,dx$

7. $\displaystyle\int \frac{(e^x - 1)^2}{e^x}\,dx$

8. $\displaystyle\int \frac{1 + e^{2x}}{e^x}\,dx$

9. $\displaystyle\int (x - 6e^{2x})\,dx$

10. $\displaystyle\int \frac{2}{x}\,dx$

11. $\displaystyle\int \frac{1}{2x}\,dx$

12. $\displaystyle\int \frac{2}{3x}\,dx$

13. $\displaystyle\int (x - 1/x)\,dx$

14. $\displaystyle\int \frac{x + 1}{x}\,dx$

15. $\displaystyle\int \frac{(x + 1)^2}{x}$

16. $\displaystyle\int \frac{(2x - 5)^2}{x}\,dx$

17. $\displaystyle\int \frac{3x^3 - 2x^2 + 1}{x^2}\,dx$

18. $\displaystyle\int \frac{1}{x^2}(2x - 3)^2\,dx$

19. Evaluate, leaving the answer in exact form.

(a) $\displaystyle\int_2^3 \frac{2}{x}\,dx$

(b) $\displaystyle\int_1^2 (e^x - 1)^2 dx$

(c) $\displaystyle\int_3^4 \frac{(x + 1)^2}{x^2}\,dx$

(d) $\displaystyle\int_{-1}^0 (e^x - 1)\,dx$

(e) $\displaystyle\int_2^3 \frac{x - 1}{x^2}\,dx$

(f) $\displaystyle\int_0^2 \frac{e^x - 1}{e^{2x}}\,dx$

(g) $\displaystyle\int_1^2 \left(\frac{1}{2x} - e^{2x}\right) dx$

(h) $\displaystyle\int_1^3 \left(\frac{2}{x^2} + \frac{3}{e^{2x}}\right) dx$

(i) $\displaystyle\int_0^1 \left(\frac{2}{\sqrt{x}} + \frac{x^{2/3}}{2} - e^{(1/2)x}\right) dx$

20. The region R in the xy plane is bounded by the curve $y = e^{-x}$, the x-axis, the y-axis and the line $x = 3$. Find the area of R in terms of e.

21. Two curves have equations $y = \dfrac{1}{x}$ and $y = x^2$.

    (a) Sketch the curves and show that they intersect where $x = 1$.

    (b) Find the area of the region bounded by these two curves, the x-axis and the line $x = 3$.

22. (a) Evaluate (i) $\displaystyle\int_{-1}^{0} (e^x - 1)\,dx$ (ii) $\displaystyle\int_{0}^{1} (e^x - 1)\,dx$ (iii) $\displaystyle\int_{-1}^{1} (e^x - 1)\,dx$

    (b) Use sketch graphs to illustrate the relationship between the results for (a).

23. Show that the equation $e^x = \dfrac{1}{x}$ has a root between 0.5 and 1.

24. Show that the equation $e^x = 2\cos x$, where x is measured in radians, has two roots in the interval $-\pi/2 < x < \pi/2$ and locate them within half a unit.

25. Explain why $\displaystyle\int_{-1}^{-2} \dfrac{1}{x}\,dx$ cannot be evaluated using $\int (1/x)\,dx = \ln x + c$.

    Show, by means of a sketch, that $\displaystyle\int_{-1}^{-2} \dfrac{1}{x}\,dx$ has a value and suggest how it can be found.

26. The diagram shows the graph of $y = x \ln x$ for the domain $1 \leqslant x \leqslant 6$.

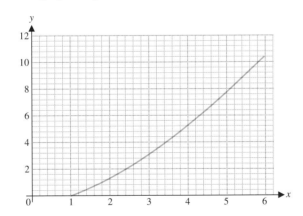

By counting squares find an approximate value for the area between the curve, the x-axis and the line $x = a$ for values of a in steps of 0.5 from $a = 1$ to $a = 6$. Plot points representing one of the curves given by $y = \int x \ln x \, dx$. Describe how a better 'picture' of the shape of this curve can be built up. If you have access to a computer with a program capable of finding numerical values for such integrals and that will also plot the points, use it to build up a picture over a larger range of values of x.

CONSOLIDATION E

SUMMARY

Integration

Integration reverses differentiation.

$$\int x^n \, \mathrm{d}x = \left(\frac{1}{n+1} \right) x^{n+1} + K$$

$$\int ax^n \, \mathrm{d}x = \left(\frac{a}{n+1} \right) x^{n+1} + K$$

Integration can be performed across a sum or difference of functions, i.e.

$$\int [\mathrm{f}(x) + \mathrm{g}(x)] \, \mathrm{d}x = \int \mathrm{f}(x) \, \mathrm{d}x + \int \mathrm{g}(x) \, \mathrm{d}x$$

Other combinations of functions can be integrated if they can be expressed as sums or differences of functions.

INTEGRATION AS A PROCESS OF SUMMATION

$$\lim_{\delta x \to 0} \sum_{x=a}^{x=b} \mathrm{f}(x) \, \delta x = \int_a^b \mathrm{f}(x) \, \mathrm{d}x$$

where $\displaystyle \int_a^b \mathrm{f}(x) \, \mathrm{d}x = \mathrm{g}(b) - \mathrm{g}(a)$, such that $\displaystyle \mathrm{g}(x) = \int \mathrm{f}(x) \, \mathrm{d}x$

Practical Applications of Calculus

AREA

The area bounded by the x-axis, the two lines $x = a$ and $x = b$, and part of the curve $y = f(x)$ can be found by summing the areas of vertical strips of width δx and using

$$\text{Area} = \lim_{\delta x \to 0} \sum_{x=a}^{x=b} y \, \delta x = \int_a^b y \, dx$$

Similarly, for horizontal strips.

$$\text{Area} = \lim_{\delta y \to 0} \sum_{y=a}^{y=b} x \, \delta y = \int_a^b x \, dy$$

Sequences

A sequence is an ordered progression of numbers which can be generated from a rule. The rule may give the rth term as a function of r, e.g. $u_r = 2r$, or it may give a relationship between successive terms together with the first term, e.g. $u_r = 2 + u_{r-1}$ and $u_1 = 1$, which is called a **recurrence** relationship.

A sequence **converges** if u_r approaches a single finite value as r increases. If a sequence does not converge it is called **divergent**. The terms in a divergent sequence can increase out of control or they can cycle through a regular pattern of values, in which case it is **periodic**, or form an irregular wave pattern and can be said to oscillate.

Number Series

Each term in a number series has a fixed numerical value.

A finite series has a finite number of terms,

e.g. $a_1 + a_2 + a_3 + \ldots + a_{10}$ is a finite series with ten terms.

The sum of the first n terms of a series is denoted by S_n, i.e.

$$S_n = a_1 + a_2 + a_3 + \ldots + a_n$$

An infinite series has no last term.

If, as $n \to \infty$, S_n tends to a finite value S, then the series converges and S is called its sum to infinity.

If $S_n \to \infty$ as $n \to \infty$ then the series is divergent.

Arithmetic Progressions

In an arithmetic progression, each term differs from the preceeding term by a constant (called the common difference).

An AP with first term a, common difference d and n terms, is

$$a,\ a+d,\ a+2d,\ldots\ \{a+(n-1)d\}$$

The sum of the first n terms of an A.P. is given by

$$S_n = \tfrac{1}{2}n(a+l) \qquad\qquad \text{where } l \text{ is the last term,}$$
$$= \tfrac{1}{2}n\{2a+(n-1)d\}$$

Geometric Progressions

In a geometric progression each term is a constant multiple of the preceeding term. This multiple is called the common ratio.

A GP with first term a, common ratio r and n terms is

$$a,\ ar,\ ar^2,\ \ldots\ ar^{n-1}$$

The sum of the first n terms is given by $S_n = \dfrac{a(1-r^n)}{1-r}$

Provided that $|r| < 1$ the sum to infinity is given by $S = \dfrac{a}{1-r}$

Exponential and Log Functions

$f(x) = a^x$ is an exponential function.

$f(x) = e^x$ where $e = 2.718\ldots$ is *the* exponential function.

When $y = e^x$, $\dfrac{dy}{dx} = e^x$, and when $y = e^{ax}$, $\dfrac{dy}{dx} = ae^{ax}$

$$\int e^x dx = e^x + k, \quad \int e^{ax} dx = \frac{1}{a}e^{ax} + k$$

The log function is the inverse of the exponential function, i.e. if $f(x) = e^x$, $f^{-1}(x)$ is denoted by $\ln x$.

When $y = \ln x$, $\dfrac{dy}{dx} = \dfrac{1}{x}$

$\Rightarrow \quad \displaystyle\int \frac{1}{x} dx = \ln x + k \ \text{ and } \ \int \frac{a}{x} dx = a\ln x + k$

TYPE I

1. $\displaystyle\int 2x(x+1)\,dx$ is

   A $x^2(x+1)+c$ C $4x+2+c$ E   2

   B $x^2(\frac{1}{2}x^2+x)+c$ D $\frac{2}{3}x^3+x^2+c$

2. $\displaystyle\int_1^2 x\,dx$ is

   A   1 B $1\frac{1}{2}$ C   3 D $-1\frac{1}{2}$ E -1

3. The sum of the series $1+5+9+13+17+21+25$ is

   A   30 B   240 C   91 D   112 E   28

4.

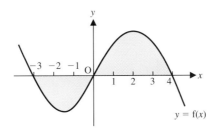

$$y = f(x)$$

   The shaded area in the diagram is given by

   A $\displaystyle\int_{-3}^4 f(x)\,dx$ C $\displaystyle\int_{-3}^1 f(x)\,dx + \int_1^4 f(x)\,dx$

   B $\displaystyle\int_{-3}^0 f(x)\,dx + \int_0^4 f(x)\,dx$ D   none of these

5. The first three terms of the series $\displaystyle\sum_{r=0}^{\infty}(-1)^{r+1}2^r x^{-r}$ are

   A $-1+\dfrac{2}{x}-\dfrac{4}{x^2}$ C $1+2x-4x^2$ E   none of these

   B $1+\dfrac{2}{x}-\dfrac{4}{x^2}$ D $\dfrac{2}{x}-\dfrac{4}{x^2}+\dfrac{8}{x^3}$

6. 3 is the arithmetic mean of a and b. Possible values of a and b are

   A   5, 4 B   0, 9 C   3, 1 D   4, 2 E   none of these

7. The series $1 - x + 2x^2 - 3x^3 + 4x^4 + \dots$ may be written more briefly as

A $\displaystyle\sum_{r=0}^{\infty} (-1)^r rx^r$ 　　　　　　　 **D** $\displaystyle 1 - \sum_{r=0}^{\infty} (-1)^r rx^r$

B $\displaystyle 1 + \sum_{r=0}^{\infty} (-1)^r rx^r$ 　　　　　 **E** $\displaystyle\sum_{r=1}^{\infty} (-1)^{r+1} rx^r$

C $\displaystyle\sum_{r=1}^{\infty} rx^r$

8. For a curve $y = f(x)$, $\dfrac{dy}{dx} > 0$ for $0 < x < a$. The graph of $y = f(x)$ could be

A

B

C

D

9. The graph of $\dfrac{dy}{dx} = f(x)$ is . The graph of $y = \displaystyle\int f(x)\,dx$ could be

A

B

C

D

E

10. The gradient at any point on the curve $y = f(x)$ is given by $\dfrac{dy}{dx} = 2x$. The point $(1, 0)$ is on the curve. The equation of the curve is

   A $y = x^2 + 1$ **C** $y = x^2 - 1$ **E** $y = 2x^2 - \frac{1}{2}$

   B $y = 2$ **D** $y = x^2$

11. $\dfrac{d}{dx}(e^{-x})$ is

   A $-x$ **B** e^{-1} **C** $-e^{-x}$ **D** $\dfrac{1}{x}$ **E** $-xe^{-x}$

12. As the sequence $u_{r+1} = \dfrac{1}{u_r}$ progresses, where $u_1 = -2$, the terms

   A converge   **B** cycle   **C** oscillate   **D** none of these

TYPE II

13. If $f(x) = (2x - 1)^2$ then $\displaystyle\int f(x)\,dx = \frac{1}{3}(x^2 - x)^3 + c$

14. If $f(x) = 3x^2$ then $\displaystyle\int_0^1 f(x)\,dx$ represents the area under $y = f(x)$ from $x = 0$ to $x = 1$

15. The sum of the first n terms, S_n, of a given series is given by $S_n = \dfrac{2n^2}{n^2 + 1}$
The first two terms of the series are $1, \frac{8}{5}$

16. The series $\displaystyle\sum_{r=2}^{12} \frac{2^r}{r}$ is a GP.

17. If $u_r = \sin\dfrac{r\pi}{2}$, $u_r \to 1$ as $r \to \infty$

18. If $2^n - 1$ is the sum of the first n terms of a series, $4^n - 1$ is the sum of the first $2n$ terms.

19. The fourth term of the series $\displaystyle\sum_{r=0}^{n} (-1)^{(r+1)} 2^r$ is 16

20. The sequence $u_n = (-1)^n$ diverges.

21. The area between the curve $y = x^2$, the y-axis and the line $y = 1$ is given
by $\displaystyle\int_0^1 x\,dy$

22. The sequence $u_r = (-1)^r \cos \pi r$ converges.

MISCELLANEOUS EXERCISE E

1. A forest fire spreads so that the number of hectares burnt after t hour is given
by $h = 30(1.65)^t$

   (a) By what constant factor is the burnt area multiplied from time $t = N$ to
   time $t = N + 1$? Express this as a percentage increase.

   (b) 1.65 can be written as e^K. Find the value of K.

   (c) Hence show that $dh/dt = 15 e^{Kt}$.

   (d) This shows that dh/dt is proportional to h. Find the constant of
   proportionality.   (OCSEB)$_s$

2. Curves C_1 and C_2 have equations $y = \dfrac{1}{x}$ and $y = kx^2$ respectively, where k
   is a constant. The curves intersect at the point P, whose x-coordinate is $\frac{1}{2}$.

   (a) Determine the value of k.

   (b) Find the gradient of C_1 at P.

   (c) Calculate the area of the finite region bounded by C_1, C_2, the x-axis and
   the line $x = 2$, giving your answer to 2 decimal places.   (ULEAC)$_s$

3. The ninth term of an arithmetic progression is 52 and the sum of the first
   twelve terms is 414. Find the first term and the common difference.   (AEB)$_s$

4. The sequence given by the iteration formula
   $$x_{n+1} = 2(1 + e^{-x_n}),$$
   with $x_1 = 0$, converges to a. Find a correct to 3 decimal places, and state an
   equation of which a is a root.   (UCLES)$_s$

5. The sequence $u_1, u_2, u_3, \ldots$, where u_1 is a given real number, is defined
   by $u_{n+1} = u_n^2 - 1$.

   (a) Describe the behaviour of the sequence for each of the cases $u_1 = 0$, $u_1 = 1$
   and $u_1 = 2$.

   (b) Given that $u_2 = u_1$, find exactly the two possible values of u_1.

   (c) Given that $u_3 = u_1$, show that
   $$u_1^4 = 2u_1^2 - u_1 = 0.$$
                                                                  (UCLES)$_s$

6. A small ball is dropped from a height of 1 m onto a horizontal floor. Each time the ball strikes the floor it rebounds to $\frac{3}{5}$ of the height from which it has just fallen.

    (a) Show that, when the ball strikes the floor for the third time, it has travelled a distance 2.92 m.

    (b) Show that the total distance travelled by the ball cannot exceed 4 m.
    (ULEAC)ₛ

7. An employer offers the following schemes of salary payments over a five year period:

    Scheme X: 60 monthly payments, starting with £1000 and increasing by £6 each month [£1000, £1006, £1012, ...];
    Scheme Y: 5 annual payments, starting with £12 000 and increasing by £d each year [£12 000, £(12 000 + d), ...].

    (a) Over the complete five year period, find the total salary payable under Scheme X.

    (b) Find the value of d which gives the same total salary for both schemes over the complete five year period.
    (ULEAC)ₛ

8. An athlete plans a training schedule which involves running 20 km in the first week of training; in each subsequent week the distance is to be increased by 10% over the previous week. Write down an expression for the distance to be covered in the nth week according to the schedule, and find in which week the athlete would first cover more than 100 km.
    (UCLES)ₛ

9. At the beginning of a month, a customer owes a credit card company £1000. In the middle of the month the customer pays £A to the company, where $A < 1000$, and at the end of the month the company adds interest at a rate of 3% of the amount still owing. This process continues, with the customer paying £A in the middle of each month, and the company adding 3% of the amount outstanding at the end of each month.

    (a) Find the value of A for which the customer still owes £1000 at the start of every month.

    (b) Find the value of A for which the whole amount owing is exactly paid off after the second payment.

    (c) Assuming that the debt has not been paid off after 4 payments, show that the amount still owing at the beginning of the 5th month can be expressed as

    $$£\{1000R^4 - A(R^4 + R^3 + R^2 + R)\}, \text{ where } R = 1.03.$$

    (d) Show that the value of A for which the whole amount owing is exactly paid off after the nth payment is given by

    $$A = \frac{1000R^{n-1}(R-1)}{R^n - 1}$$
    (UCLES)ₛ

10. For the infinite geometric series $1 + 2x + 4x^2 + 8x^3 + \ldots$,

   (a)   state the condition that the series has a sum to infinity,

   (b)   find the sum to infinity of the series, assuming the sum exists.

   The sum to infinity is to be approximated by taking the first ten terms of the series.

   (c)   Find the sum of the first ten terms.

   (d)   Show that the error in this approximation is $\dfrac{1024x^{10}}{1 - 2x}$.

   (e)   For what values of x will this error be less than one hundredth of the sum to infinity?   (MEI)$_s$

11. Given that $y = x(x + 2)$ find   (a) $\dfrac{dy}{dx}$ (b) $\displaystyle\int y\,dx$

12.

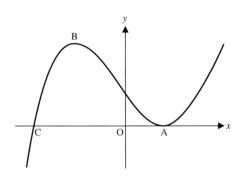

The figure shows a sketch of part of the curve with equation

$$y = x^3 - 12x + 16$$

Calculate

   (a)   the coordinates of the turning points A and B

   (b)   the coordinates of C, the point where the curve crosses the x-axis

   (c)   the area of the finite region enclosed by the curve and the x-axis. (ULEAC)

13. (a)   Evaluate $\displaystyle\int_1^4 x(x - 1)\,dx$

   (b)   Indicate on a sketch the region whose area is equal to the definite integral in part (a)

14. A curve whose equation is $y = f(x)$ passes through the point $(2, 1)$. The gradient function is given by $f'(x) = \frac{1}{2}(1 - 2x)^2$. Find the equation of the curve.

15. Evaluate (a) $\displaystyle\int_{-1}^{0}(x-1)^3\,dx$ (b) $\displaystyle\int_{0}^{\pi}\pi x^2\,dx$

16. An arithmetic progression has first term a and common difference -1. The sum of the first n terms is equal to the sum of the first $3n$ terms. Express a in terms of n.   (UCLES)

17. Find the positive constants a and b such that 0.25, a, 9 are in geometric progression and 0.25, a, $9-b$ are in arithmetic progression.   (ULEAC)

18. Sketch the curve $y=(x-1)^2(x-2)$, showing the coordinates of the points where it meets the axes and of the turning points. Calculate the area enclosed by the curve and the x-axis.

19. Given that $f(x)=x(x-3)^2$, find the two values of x for which $f'(x)=0$. Give the corresponding values of $f(x)$ and show that one of these is a maximum and the other is a minimum.
Sketch the curve $y=f(x)$.
Find the area of the region enclosed by the curve and the x-axis between $x=0$ and $x=3$

20. Sketch in separate diagrams the curves whose equations are

$$y = \ln x, \ x>0$$
$$y = |\ln x|, \ x>0$$

Calculate, correct to two decimal places, the values of x for which

$$|\ln x| = 2$$

Giving a reason, state the number of real roots of the equation

$$|\ln x| + x = 3$$ (AEB)

21. (a) Sketch the curve $y=1-e^x$

    (b) Find the area enclosed by the curve $y=1-e^x$, the x-axis and the line $x=2$.

22. Find the area of the region enclosed by the curve $y=1+\dfrac{1}{x}$, the line $y=2x$ and the line $x=4$.

23. Evaluate (a) $\displaystyle\int_{1}^{2}\frac{x+1}{x}\,dx$ (b) $\displaystyle\int_{1}^{2}(e^{2x}-1)\,dx$

24. The diagram shows the area enclosed by the curve $y = x^2 + 1$ and the line $y = 3x + 5$.

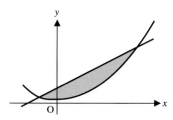

(a) Find the coordinates of the points of intersection of $y = x^2 + 1$ and $y = 3x + 5$.

(b) Find the area shaded in the diagram.

ANSWERS

Answers to questions taken from past examination papers are the sole responsibility of the authors and have not been approved by the Examining Boards.

CHAPTER 1
Exercise 1a – p. 2
1. $15x$
2. $2x^2$
3. $4x^2$
4. $10pq$
5. $8x^2$
6. $10p^2qr$
7. $9a^2$
8. $63ab$
9. $24st^2$
10. $8a^3$
11. $4ab^3$
12. $196p^4q^2$
13. $12a^2b^3$
14. $100x^2y^2$
15. $12a^4b^2$

Exercise 1b – p. 3
1. $3x^2 - 4x$
2. $a - 12$
3. $2y - xy + y^2$
4. $5pq - 9p^2$
5. $3xy + y^2$
6. $x^3 - x^2 + x + 7$
7. $5 + t - t^2$
8. $a^2 - ab - 2b$
9. $7 - x$
10. $4x - 9$
11. $3x^2 + 18x - 20$
12. $ab - 2ac + cb$
13. $11cT - 2cT^2 - 55T^2$
14. $-x^3 + 7x^2 - 7x$
15. $-4y^2 + 24y - 10$
16. $5RS + 5RF - R^2$

Exercise 1c – p. 4
1. $\dfrac{a}{3}$
2. $\dfrac{2}{a}$
3. $\dfrac{q}{r}$
4. $\dfrac{2}{3b}$

5. $\dfrac{3p}{8qr}$
6. $2a$
7. $6ax$
8. $2x$
9. $\dfrac{9b}{5a}$
10. $\frac{5}{3}x$
11. $2m$
12. $5xy$

Exercise 1d – p. 5
1. $\dfrac{7}{4x}$
2. $\dfrac{3}{5}x$
3. (a) False   (b) False   (c) True
4. $\frac{2}{5}x$
5. x^3/y^2
6. $\dfrac{2x^2}{3y^2}$
7. $\dfrac{6t^2}{s}$
8. $\dfrac{8v^2}{3}$
9. $\dfrac{2r}{3}$
10. $3x$
11. $\dfrac{9x}{4y^2}$
12. $\dfrac{x^2}{24}$
13. $\dfrac{2}{a(a+b)}$
14. $\dfrac{1}{2a}$
15. $\dfrac{a^4}{27}$
16. $\dfrac{3}{2(x+3)}$
17. $\frac{2}{3}$

18. $\dfrac{2r}{3s^2}$

19. $\dfrac{3x^2}{2(y-2)}$

20. $\dfrac{b^2}{c^2}$

21. $2a^3/3$

Exercise 1e – p. 6
1. -7
2. 2
3. (a) 1   (b) -5 (c) -1
4. (a) 1   (b) 0   (c) -3

Exercise 1f – p. 7
1. $x^2 + 6x + 8$
2. $x^2 + 8x + 15$
3. $a^2 + 13a + 42$
4. $t^2 + 15t + 56$
5. $s^2 + 17s + 66$
6. $2x^2 + 11x + 5$
7. $5y^2 + 28y + 15$
8. $6a^2 + 17a + 12$
9. $35t^2 + 86t + 48$
10. $99s^2 + 49s + 6$
11. $x^2 - 5x + 6$
12. $y^2 - 5y + 4$
13. $a^2 - 11a + 24$
14. $b^2 - 17b + 72$
15. $p^2 - 15p + 36$
16. $2y^2 - 13y + 15$
17. $3x^2 - 13x + 4$
18. $6r^2 - 25r + 14$
19. $20x^2 - 19x + 3$
20. $6a^2 - 7ab + 2b^2$
21. $x^2 - x - 6$
22. $a^2 + a - 56$
23. $y^2 + 2y - 63$
24. $s^2 + s - 30$
25. $q^2 + 8q - 65$
26. $2t^2 + 3t - 20$
27. $4x^2 + 11x - 3$
28. $6q^2 - q - 15$
29. $x^2 - xy - 2y^2$
30. $2s^2 + st - 6t^2$

Exercise 1g – p. 7
1. $x^2 - 4$
2. $25 - x^2$
3. $x^2 - 9$
4. $4x^2 - 1$
5. $x^2 - 64$
6. $x^2 - a^2$
7. $x^2 - 1$
8. $9b^2 - 16$
9. $4y^2 - 9$
10. $a^2b^2 - 36$
11. $25x^2 - 1$
12. $x^2y^2 - 16$

Exercise 1h – p. 8
1. $x^2 + 8x + 16$
2. $x^2 + 4x + 4$
3. $4x^2 + 4x + 1$
4. $9x^2 + 30x + 25$
5. $4x^2 + 28x + 49$
6. $x^2 - 2x + 1$
7. $x^2 - 6x + 9$
8. $4x^2 - 4x + 1$
9. $16x^2 - 24x + 9$
10. $25x^2 - 20x + 4$
11. $9t^2 - 42t + 49$
12. $x^2 + 2xy + y^2$
13. $4p^2 + 36p + 81$
14. $9q^2 - 66q + 121$
15. $4x^2 - 20xy + 25y^2$

Exercise 1i – p. 9
1. $11x - 2x^2 - 12$
2. $x^2 - 49$
3. $6 - 25x + 4x^2$
4. $14p^2 - 3p - 2$
5. $9p^2 - 6p + 1$
6. $15t^2 + t - 2$
7. $16 - 8p + p^2$
8. $14t - 3 - 8t^2$
9. $x^2 + 4xy + 4y^2$
10. $16x^2 - 9$
11. $9x^2 + 42x + 49$
12. $15 - R - 2R^2$
13. $a^2 - 6ab + 9b^2$
14. $4x^2 - 20x + 25$
15. $49a^2 - 4b^2$
16. $9a^2 + 30ab + 25b^2$
17. (a) 6, -22 (b) 15, 31
       (c) 14, -31 (d) 81, 18

Exercise 1j – p. 11
1. $(x+5)(x+3)$
2. $(x+7)(x+4)$
3. $(x+6)(x+1)$
4. $(x+4)(x+3)$
5. $(x-1)(x-9)$
6. $(x-3)^2$
7. $(x+6)(x+2)$
8. $(x-8)(x-1)$

9. $(x+7)(x-2)$
10. $(x+4)(x-3)$
11. $(x-5)(x+1)$
12. $(x-12)(x+2)$
13. $(x+7)(x+2)$
14. $(x-1)^2$
15. $(x-3)(x+3)$
16. $(x+8)(x-3)$
17. $(x+2)^2$
18. $(x-1)(x+1)$
19. $(x-6)(x+3)$
20. $(x+5)^2$
21. $(x-4)(x+4)$
22. $(4+x)(1+x)$
23. $(2x-1)(x-1)$
24. $(3x+1)(x+1)$
25. $(3x-1)^2$
26. $(3x+1)(2x-1)$
27. $(3+x)^2$
28. $(2x-3)(2x+3)$
29. $(x+a)^2$
30. $(xy-1)^2$

Exercise 1k – p. 12

1. $(3x-4)(2x+3)$
2. $(4x-3)(x-2)$
3. $(4x-1)(x+1)$
4. $(3x-2)(x-5)$
5. $(2x-3)^2$
6. $(1-2x)(3+x)$
7. $(5x-4)(5x+4)$
8. $(3+x)(1-x)$
9. $(5x-1)(x-12)$
10. $(3x+5)^2$
11. $(3-x)(1+x)$
12. $(3+4x)(4-3x)$
13. $(1+x)(1-x)$
14. $(3x+2)^2$
15. $(x+y)^2$
16. $(1-2x)(1+2x)$
17. $(2x-y)^2$
18. $(3-2x)(3+2x)$
19. $(6+x)^2$
20. $(5x-4)(8x+3)$
21. $(7x+30)(x-5)$
22. $(6-5x)(6+5x)$
23. $(x-y)(x+y)$
24. $(9x-2y)^2$
25. $(7-6x)^2$
26. $(5x-2y)(5x+2y)$
27. $(6x+5y)^2$
28. $(2x-3y)(2x+y)$
29. $(3x+4y)(2x+y)$
30. $(7pq-2)^2$

Exercise 1l – p. 13

1. not possible
2. $2(x+1)^2$
3. $(x+2)(x+1)$
4. $3(x+5)(x-1)$
5. not possible
6. not possible
7. not possible
8. $2(x-2)^2$
9. $3(x-2)(x+1)$
10. $2(x^2-3x+4)$
11. $3(x-4)(x+2)$
12. $(x-6)(x+2)$
13. not possible
14. $4(x-5)(x+5)$
15. $5(x^2-5)$
16. not possible
17. not possible
18. not possible
19. $x(x-3)(x+1)$
20. $x(x-2)(x+2)$
21. $y(x^2+2x+2)$
22. $x(x^2+1)$
23. $xy(xy-1)$
24. $pq(q-p)(q+p)$

Exercise 1m – p. 14

1. $\frac{1}{4}$
2. $\frac{2(x+2)}{3(x-2)}$
3. $\frac{2}{3}$
4. $\frac{3}{5}$
5. $\frac{x}{y}$
6. not possible
7. not possible
8. $\frac{5x(x+y)}{5y+2x}$
9. $\frac{2a-6b}{6a+b}$
10. $\frac{b-4}{3x(b+4)}$
11. $\frac{x-3}{x+4}$
12. $\frac{4y^2+3}{(y+3)(y-3)}$
13. $\frac{1}{3(x+3)}$
14. $\frac{x+2}{2x+1}$

15. $\dfrac{x-2}{x-1}$

16. $\dfrac{1}{2(a-5)}$

17. $\dfrac{3}{p+3q}$

18. $\dfrac{a^2+2a+4}{(a+5)(a+2)}$; not possible

19. $\dfrac{x+1}{3(x+3)}$

20. $\dfrac{4(x-3)}{(x+1)^2}$

21. $\dfrac{1}{x+1}$

22. $\dfrac{1}{x-1}$

23. $2(x+3)$

24. $\dfrac{x(2x-3)}{x-1}$

25. $\dfrac{x+2}{2x(x-2)}$

26. $\dfrac{x^2}{x-1}$

Exercise 1n – p. 16

1. x^3-x^2-x-2
2. $3x^3-5x^2-x+2$
3. $4x^3-8x^2+13x-5$
4. x^3-2x^2+1
5. $2x^3-9x^2-24x-9$
6. $x^3+6x^2+11x+6$
7. x^3+4x^2-x-4
8. x^3-4x^2+x+6
9. $2x^3+7x^2+7x+2$
10. x^3+4x^2+5x+2
11. $4x^3+4x^2-7x+2$
12. $27x^3-27x^2+9x-1$
13. $4x^3-9x^2-25x-12$
14. $4x^3-4x^2-x+1$
15. $6x^3+13x^2+x-2$
16. x^3+3x^2+3x+1
17. x^3+x^2-4x-4
18. $2x^3+7x^2-9$
19. $24x^3+38x^2-51x+10$
20. $4x^3-42x^2+68x+210$
21. $3x^3-16x^2+28x-16;\ -16,\ 28$
22. $6,\ -17$
23. $x^3+3x^2y+3xy^2+y^3$
24. $x^4+4x^3y+6x^2y^2+4xy^3+y^4$

Exercise 1o – p. 18

1. $x^3+9x^2+27x+27$
2. $x^4-8x^3+24x^2-32x+16$
3. $x^4+4x^3+6x^2+4x+1$
4. $8x^3+12x^2+6x+1$
5. $x^5-15x^4+90x^3-270x^2+405x-243$
6. $p^4-4p^3q+6p^2q^2-4pq^3+q^4$
7. $8x^3+36x^2+54x+27$
8. $x^5-20x^4+160x^3-640x^2$
$+1280x-1024$
9. $81x^4-108x^3+54x^2-12x+1$
10. $1+20a+150a^2+500a^3+625a^4$
11. $64a^6-192a^5b+240a^4b^2-160a^3b^3$
$+60a^2b^4-12ab^5+b^6$
12. $8x^3-60x^2+150x-125$

Mixed Exercise 1 – p. 18

1. -23
2. $115x-105x^2-30$
3. 108
4. $3(x-2)(x-1)$
5. 250
6. $4(x-3)(x+3)$
7. $4x^3-4x^2+x$
8. $(x-5)^2$
9. 7
10. $12x^3-3x$
11. $(2x-3y)^2$
12. 7
13. -12
14. $(3x+2y)(2x-5y)$
15. 24
16. (a) $\dfrac{x+3}{2}$ (b) $\dfrac{1}{x-3}$

17. $\dfrac{2}{3(n+2)}$

18. $\dfrac{x+3}{x-2}$

19. $(x-1)^2$

20. $\dfrac{(x-1)^3}{x+1}$

21. $(x-4)(x-1)$
22. $(x+1)^2$
23. $y(x-1)(x+1)$
24. (a) $(2-x)(5+x)$
(b) $x(x-6)(x+2)$
25. (a) $(x-2)^2$
(b) $(x-2)^2$
(c) 2 sq. units
(d) $\dfrac{(x-2)^2}{x^2-2}$

CHAPTER 2

Exercise 2a – p. 23
1. (a) 1 m   (b) 2 m   (c) 2 m
2. (a) (i) 0.5 mm   (ii) 1 mm
    (b) (i) 1 mm   (ii) 2 mm
3. (a) (i) $1.15 \leqslant x < 1.25$
       (ii) $3.45 \leqslant y < 3.55$
       (iii) $3.9675 \leqslant xy < 4.4375$
       (iv) $0.3239\ldots \leqslant \dfrac{x}{y} < 0.3623$
    (b) (i) -5.4% to 5.9% (2 sf)
       (ii) -5.4% to 5.9% (2 sf)
4. (a) (i) $1.6 \leqslant x < 1.7$
       (ii) $0.5 \leqslant y < 0.6$
       (iii) $0.8 \leqslant xy < 1.02$
       (iv) $2.6 \leqslant \dfrac{x}{y} < 3$
    (b) (i) -22% to 0 (ii) -20% to 5.9%
5. (a) 0.284% (b) 4
6. 13.5 m (3 sf)

Exercise 2b – p. 26
1. $2\sqrt{3}$
2. $4\sqrt{2}$
3. $3\sqrt{3}$
4. $5\sqrt{2}$
5. $10\sqrt{2}$
6. $6\sqrt{2}$
7. $9\sqrt{2}$
8. $12\sqrt{2}$
9. $5\sqrt{3}$
10. $4\sqrt{3}$
11. $10\sqrt{5}$
12. $2\sqrt{5}$

Exercise 2c – p. 27
1. $2\sqrt{3} - 3$
2. $5\sqrt{2} + 8$
3. $2\sqrt{5} + 5\sqrt{15}$
4. 4
5. $\sqrt{6} + \sqrt{2} - \sqrt{3} - 1$
6. $13 + 7\sqrt{3}$
7. 4
8. $5 - 3\sqrt{2}$
9. $22 - 10\sqrt{5}$
10. 9
11. $10 - 4\sqrt{6}$
12. $31 + 12\sqrt{3}$
13. $(1 - \sqrt{2})$
14. $(1 + \sqrt{3})$
15. $(\sqrt{2} + 2)$
16. $(\sqrt{5} - 1)$
17. $(4 + \sqrt{5})$
18. $(\sqrt{11} - 3)$
19. $(2\sqrt{3} + 4)$
20. $(\sqrt{6} + \sqrt{5})$
21. $(3 + 2\sqrt{3})$
22. $(2\sqrt{5} + \sqrt{2})$

Exercise 2d – p. 28
1. $\frac{3}{2}\sqrt{2}$
2. $\frac{1}{7}\sqrt{7}$
3. $\frac{2}{11}\sqrt{11}$
4. $\frac{3}{5}\sqrt{10}$
5. $\frac{1}{9}\sqrt{3}$
6. $\frac{1}{2}\sqrt{2}$
7. $\sqrt{2} + 1$
8. $\frac{1}{23}(15\sqrt{2} - 6)$
9. $\frac{1}{3}(4\sqrt{3} + 6)$
10. $-5(2 + \sqrt{5})$
11. $\frac{1}{4}(\sqrt{7} + \sqrt{3})$
12. $4(2 + \sqrt{3})$
13. $\sqrt{5} - 2$
14. $\frac{1}{13}(7\sqrt{3} + 2)$
15. $3 + \sqrt{5}$
16. $3(\sqrt{3} + \sqrt{2})$
17. $\frac{3}{19}(10 - \sqrt{5})$
18. $3 + 2\sqrt{2}$
19. $\frac{2}{3}(7 - 2\sqrt{7})$
20. $\frac{1}{2}(1 + \sqrt{5})$
21. $\frac{1}{4}(\sqrt{11} + \sqrt{7})$
22. $\frac{1}{6}(9 + \sqrt{3})$
23. $\frac{1}{14}(9\sqrt{2} - 20)$
24. $\frac{1}{6}(3\sqrt{2} + 2\sqrt{3})$
25. $\frac{1}{2}(2 + \sqrt{2})$
26. $\frac{1}{42}(3\sqrt{7} - \sqrt{21})$
27. $\frac{1}{9}(\sqrt{30} + 2\sqrt{3})$

Exercise 2e – p. 32
1. 2^6
2. x^8
3. $a^3 b^6$
4. a^{-2}
5. $6x^5$
6. 8×3^5
7. $\frac{1}{2^2}$
8. $\frac{1}{4^2}$

9. $\dfrac{1}{x^6}$

10. $\dfrac{1}{2^4}$

11. $\dfrac{1}{2^2}$

12. 3^2

13. x^2

14. 1

15. t^4

16. 1

17. 2

18. $y^{3/2}$

19. x^5

20. $\dfrac{1}{y^{3/4}}$

21. p

22. $\frac{1}{2}$

23. 9

24. 27

25. 8

26. 9

27. $\frac{16}{9}$

28. 4

29. $\frac{125}{8}$

30. $\frac{1}{2}$

31. 3

32. $\frac{1}{32}$

33. $\frac{1}{2}$

34. 2

35. 27

36. $\frac{9}{4}$

37. 1

38. 16

39. $\frac{5}{4}$

40. -5

41. 1331

42. $\frac{3}{5}$

43. 6

44. $\frac{16}{27}$

45. 8

46. 5

47. 1

48. 1

Mixed Exercise 2 – p. 33

1. (a) $2\sqrt{21}$ (b) $10\sqrt{3}$ (c) $3\sqrt{5}$
(d) $3\sqrt{3}$ (e) $5\sqrt{10}$ (f) $3\sqrt{2}$

2. (a) $\sqrt{3}-3$ (b) $3\sqrt{2}-4$
(c) $7\sqrt{3}-6$ (d) $7-2\sqrt{10}$

3. (a) $8-2\sqrt{2}$ (b) $7-2\sqrt{10}$

4. (a) $(7+\sqrt{3})(7-\sqrt{3})=46$
(b) $(2\sqrt{2}-1)(2\sqrt{2}+1)=7$
(c) $(\sqrt{7}+\sqrt{5})(\sqrt{7}-\sqrt{5})=2$

5. (a) $\frac{5}{7}\sqrt{7}$ (b) $\frac{1}{3}(\sqrt{13}+2)$
(c) $4(\sqrt{3}+\sqrt{2})$ (d) $2-\sqrt{3}$

6. (a) 32 (b) 27 (c) 3   (d) $\frac{1}{4}$ (e) 15

7. (a) 1   (b) 1

8. (a) $\frac{1}{4}$ (b) $\frac{4}{7}$ (c) $\frac{16}{243}\sqrt{6}$

9. (a) 1   (b) $\frac{1}{5}\sqrt{15}$

10. (a) 3   (b) $-\frac{3}{4}$ (c) 4

11. (a) 0.845, 0.472, 1.10   (b) 0.0731
(c) $\frac{1}{3}\sqrt{3}(\sqrt{3}-1)^2(\sqrt{5}-2)$
(d) 0.0730   (e) 0.14% (2 sf)

CHAPTER 3

Exercise 3a – p. 39

1. 7

2. 2

3. 2

4. -2

5. 4

6. 1

7. -10

8. 3/4

9. 25/8

10. -1

11. 32/63

12. 5

13. 4/3

14. $-2/3$

15. 5/14

16. $(4-b)/a$

17. $\frac{3}{2}(c-3y)$

18. $-(c+by)/a$

Exercise 3b – p. 40

1. $x=-2$ or $x=-3$

2. $x=2$ or $x=-3$

3. $x=3$ or $x=-2$

4. $x=-2$ or $x=-4$

5. $x=1$ or $x=3$

6. $x=1$ or $x=-3$

7. $x=-1$ or $x=-\frac{1}{2}$

8. $x = 2$ or $x = \frac{1}{4}$

9. $x = 1$ or $x = -5$

10. $x = 8$ or $x = -9$

11. $-1, 3$

12. $-1, -4$

13. $1, 5$

14. $2, -5$

15. $-2, 7$

16. $2, 7$

Exercise 3c – p. 42

1. $x = 2$ or $x = 5$

2. $x = 3$ or $x = -5$

3. $x = 4$ or $x = -1$

4. $x = 3$ or $x = 4$

5. $x = \frac{1}{3}$ or $x = -1$

6. $x = -1$ or $x = -6$

7. $x = 0$ or $x = 2$

8. $x = -1$ or $x = -\frac{1}{4}$

9. $x = \frac{2}{3}$ or $x = -1$

10. $x = 0$ or $x = -\frac{1}{2}$

11. $x = 0$ or $x = -6$

12. $x = 0$ or $x = 10$

13. $x = 0$ or $x = \frac{1}{2}$

14. $x = 5$ or $x = -4$

15. $x = 2$ or $x = -\frac{4}{3}$

16. $x = 2$ or $x = -1$

17. $x = 0$ or $x = 1$

18. $x = 0$ or $x = 2$

19. $x = 3$ or $x = -1$

20. $x = -1$ or $x = \frac{1}{2}$

Exercise 3d – p. 44

1. 4

2. 1

3. 9

4. 25

5. 2

6. $\frac{25}{4}$

7. 192

8. 81

9. 200

10. $\frac{1}{4}$

11. $\frac{1}{3}$

12. $\frac{9}{8}$

13. $x = -4 \pm \sqrt{17}$

14. $x = 1 \pm \sqrt{3}$

15. $x = -\frac{1}{2}(1 \pm \sqrt{5})$

16. $x = -\frac{1}{2}(1 \pm \sqrt{3})$

17. $x = -\frac{1}{2}(3 \pm \sqrt{5})$

18. $x = \frac{1}{4}(1 \pm \sqrt{17})$

19. $x = -2 \pm \sqrt{6}$

20. $x = -\frac{1}{6}(1 \pm \sqrt{13})$

21. $x = \frac{1}{2}(-2 \pm 3\sqrt{2})$

22. $x = \frac{1}{2}(1 \pm \sqrt{13})$

23. $x = -\frac{1}{8}(1 \pm \sqrt{17})$

24. $x = -\frac{1}{4}(3 \pm \sqrt{41})$

Exercise 3e – p. 46

1. $x = -2 \pm \sqrt{2}$

2. $x = \frac{1}{4}(-1 \pm \sqrt{17})$

3. $x = \frac{1}{2}(-5 \pm \sqrt{21})$

4. $x = \frac{1}{4}(1 \pm \sqrt{33})$

5. $x = 2 \pm \sqrt{3}$

6. $x = \frac{1}{4}(1 \pm \sqrt{41})$

7. $x = \frac{1}{6}(1 \pm \sqrt{13})$

8. $x = -\frac{1}{6}(1 \pm \sqrt{13})$

9. $x = 1 \pm \sqrt{6}$

10. $x = \frac{1}{5}(1 \pm \sqrt{21})$

11. $x = -0.260$ or -1.540

12. $x = 2.781$ or 0.719

13. $x = 1.883$ or -0.133

14. $x = 0.804$ or -1.554

15. $x = 0.804$ or -1.554

16. $x = 0.724$ or 0.276

17. $x = 3.303$ or -0.303

18. $x = 7.873$ or 0.127

19. $x = 1.281$ or -0.781

20. $x = 7.873, 0.127$

Exercise 3f – p. 48

1. $x = 3, y = -1$

2. $x = -1, y = -2$

3. $x = 10/3, y = 1$

4. $x = 3, y = 1$

5. $x = \frac{1}{2}(a + b), y = \frac{1}{2}(a - b)$

6. $x = -1, y = 5$

7. $x = 8/3, y = 1/3$

8. $x = 2, y = 4$

9. $x = -1 - \sqrt{6}, y = \sqrt{3} + \sqrt{2}$

10. $x = (2 + 5b)/(a + b),$
$y = (2 - 5a)/(a + b)$

11. $x = (a + c)/(1 - m),$
$y = (c + am)/(1 - m)$

12. $x = 2(c - t)/t, y = c - 2t$

Exercise 3g – p. 49

1.

x	-2	1
y	-1	2

2. $x = -1, y = 3$

3. $x = 2, y = 3$

4.

x	-1	$\frac{1}{2}$
y	4	2

5.

x	2	$-\frac{1}{2}$
y	-3	2

6.

x	$\frac{7}{2}$	-2
y	$-\frac{1}{2}$	5

7.

x	1	2
y	2	1

8.

x	-1	3
y	-4	4

9. $x = 1, y = 5$

10.

x	6	-6
y	2	-4

11. $x = \frac{1}{2}, y = -1$

12.

x	1	0
y	$\frac{1}{3}$	$\frac{2}{3}$

13. $x = -1, y = -\frac{1}{2}$

14. $x = 1, y = -\frac{1}{3}$

15.

x	$-\frac{1}{3}$	$\frac{2}{3}$
y	$-\frac{1}{2}$	$\frac{1}{4}$

16. $x = 1, y = \frac{1}{2}$

17.

x	-3	6
y	-3	$\frac{3}{2}$

18.

x	1	$-\frac{1}{4}$
y	-2	3

19.

x	-1	2
y	$\frac{1}{3}$	$-\frac{1}{6}$

20.

x	$\frac{1}{2}$	0
y	$\frac{1}{2}$	1

21.

x	1	$3\frac{1}{2}$
y	1	-4

Exercise 3h – p. 53

1. 4

2. $-\frac{5}{3}$

3. 1

4. $\frac{4}{3}$

5. -3

6. $\frac{2}{5}$

7. real and different

8. not real

9. real and different

10. real and equal

11. real and different

12. real and equal

13. real and different

14. not real

15. real and different

16. real and equal

17. $k = \pm 12$

18. $a = 2\frac{1}{4}$

19. $p = 2$

22. $q^2 = 4p$

Mixed Exercise 3 – p. 54

1. $\frac{3}{2}$

2. $5/4$

3. $-3/2$

4. $\sqrt{3} + \sqrt{2}$

5. $(3 - \sqrt{2})/(\sqrt{2} - 1)$

6. $(b + c)/(2 - a)$

7. $x = 2, y = 0$

8. $x = 3/2, y = 7/4$

9. $x = 34/7, y = -5/7$

10. $x = -\sqrt{2}, y = -(2 + \sqrt{2})$

11. $x = (3 - 2b)/(2a + 1)$,
$y = (3a + b)/(2a + 1)$

12. $x = (2b + ac)/(a + b)$,
$y = (c - 2)/(a + b)$

13. (a) 5   (b) $-1, 6$ (c) 5

14. (a) 6   (b) $3 \pm \sqrt{14}$ (c) 6

15. (a) $-\frac{3}{2}$ (b) $\frac{1}{4}(-3 \pm \sqrt{17})$ (c) $-\frac{3}{2}$

16. (a) $-\frac{4}{3}$ (b) $\frac{1}{3}(-2\pm\sqrt{19})$ (c) $-\frac{4}{3}$

17. (a) 2 (b) 1, 1 (c) 2

18. (a) $\frac{11}{4}$ (b) $-\frac{1}{4}$, 3 (c) $\frac{11}{4}$

19. (a) -1 (b) $\frac{1}{2}(-1\pm\sqrt{13})$ (c) -1

20. (a) -4 (b) -6, 2 (c) -4

21. (a) -2 (b) $-1\pm\sqrt{3}$ (c) -2

22. (a) -4 (b) -2, -2 (c) -4

23. $x = 0$ or 2

24. $x = -4$ or 1

25. $x = 2$ or $x \to \infty$

27. $x = \frac{1}{2}(7\pm\sqrt{89})$

28. $x = 1$ or $x \to \infty$

29.

x	4	-22
y	5	31

30.

x	2	4
y	5	9

31. $x = 5$ or $\frac{2}{3}$; (a) rational
(b) it factorises

32. (a) not real (b) real and different
(c) real and equal
(d) real and different

33. 4, -1

35. 2

36. line 2: dividing by x loses the solution
$x = 0$

line 3: $\dfrac{1}{x-1}\times 2 = \dfrac{2}{x-1}$ *not* $\dfrac{2}{2x-2}$

line 4: $\dfrac{2}{2x-2}= x \Rightarrow 2 = 2x^2 - 2x$

not $2 = 2x^2 - 2$
correct solution: $x = -1$, 0 or 2

CONSOLIDATION A

Multiple Choice Exercise A – p. 57
1. A
2. E
3. E
4. B
5. E
6. A
7. E
8. C
9. F
10. F
11. T
12. T
13. T
14. T
15. F
16. T
17. F

Miscellaneous Exercise A – p. 58
1. (a) 3/2 (b) 7/2
2. $(2x-5)(x+1)$
3. $(3, -1)$. $(-1/3, 7/3)$
4. -4
5. -5 or 4
6. $3[(x+1)^2 - 7/3]$
7. $p = 8$, $q = 2$
8. $x = -3$, $y = 0$; $x = \dfrac{9}{5}$, $y = \frac{12}{5}$
9. 81/4
10. 1
12. $k < 0$, $k > 3$
13. (a) $x(x^2+7)$ (b) $(x-y)(x+y)$
(c) $(a-2-b)(a-2+b)$
14. $-5\sqrt{5} - 9$
15. 16/3
16. $x = 2$, $y = 2$
17. $p = -4$, $q = 1/2$
18. 1
19. (a) $11\sqrt{3} - 7$ (b) $5x^2 + x - 2$
20. $x = 2.35$, $y = 0.10$; $x = -12.35$, $y = 9.90$
21. $a = 3$, $b = -7$, $c = 17$
22. $x = -1$, 0, 2
23. (a) $(x-5)/2(x+1)$
(b) $\frac{1}{2}(9\sqrt{7} - 7)$
24. $a = 3$, $b = -36$
26. (a) 10% (b) 16.5%
27. (a) $a^3b - ac^2$ (b) $a(1-b)/c^2$
28. (a) $x = (-5\pm5\sqrt{1+4\pi})/2\pi$
(b) $9\pi^3/(1+9\pi^3)$
29. (a) 7.69 (b) 7.71 (c) -0.26%
30. $S = m_2(3m_1{}^2 + m_2)/(2m_2 - m_1)$,
$T = m_2(1+6m_1)/(2m_2 - m_1)$
31. (a) 3/2 (b) 7 (c) 2/3 (d) 0
32. $x = -2$

CHAPTER 4

Exercise 4a – p. 65
1. 2 cm
2. 90 cm
3. 20 cm
4. (a) 84 cm (b) $7x$ cm
5. 5 : 7
6. 3 : 1
7. $y : (x-y)$

8. $\dfrac{ma}{n-m}$

9. $(a+b):b$

Exercise 4b – p. 67

1. 3.75 cm

2. (a) TN = 2.5 cm   (b) LN = 4.5 cm

3. BC = 2.25 cm

4. $\dfrac{yz}{x}$

5. (a) 1.25 cm   (b) 2 : 7 externally

6. (a) Y   (b) Y   (c) Y
   (d) Y   (e) N   (f) Y

Exercise 4c – p. 71

1. congruent   **2.** similar   **3.** neither

4. congruent   **5.** congruent   **6.** congruent

7. neither   **8.** similar   **9.** neither

Exercise 4d – p. 73

2. $\frac{5}{26}$ cm

3. XZ = $\frac{7}{4}$ cm, QR = 16 cm

8. RT = 12 cm, AT = 8 cm

9. 9 : 4

10. 10°

11. BD = $\frac{24}{7}$, DC = $\frac{25}{7}$, AD = $\frac{120}{7}\sqrt{2}$

13. EC = 3 cm, AC = $\frac{25}{3}$ cm

CHAPTER 5

Exercise 5a – p. 78

1.

2. $(9,5)$ and $(9,1)$ or $(-3,5)$ and $(-3,-1)$

3. $(-2,2), (3,-3)$

Exercise 5b – p. 81

1. (a) 5   (b) $\sqrt{2}$ (c) $\sqrt{13}$

2. (a) $\sqrt{109}$ (b) $\sqrt{5}$ (c) $2\sqrt{2}$

3. $\sqrt{65}$

4. $\sqrt{13}$

5. $4\sqrt{5}$

7. (b) $17\frac{1}{2}$ sq units

8. $\sqrt{5}(2+\sqrt{2})$

9. $(-5, -3)$

11. $(6, 6)$

12. $(1, -1)$

13. The x coordinate of D is
   $\frac{1}{2}$ (x coord. A + x coord. B), yes

14. $\left(\dfrac{a+c}{2}, \dfrac{b+d}{2}\right)$

Exercise 5c – p. 86

1. (a) 3   (b) $\frac{3}{2}$ (c) $\frac{1}{3}$ (d) $\frac{3}{4}$
   (e) -4 (f) 6   (g) $-\frac{7}{3}$
   (h) $-\frac{3}{2}$ (i) $-\frac{5}{3}$ (j) $\dfrac{k}{h}$

2. (a) yes   (b) no   (c) yes
   (d) yes   (e) no

3. (a) parallel   (b) perpendicular
   (c) perpendicular   (d) neither
   (e) parallel   (f) neither

Exercise 5d – p. 87

1. $a = 0, b = 4$

2. (b) $22\frac{1}{2}$ square units

5. $\sqrt{(a^2 + 4b^2)}$

9. $(a-2)^2 + (b-1)^2 = 9$

10. $b(d-b) = ac$

11. $b^2 = 8a - 16$

13. $(1, 0), (0, -1), \frac{1}{2}$ square unit

CHAPTER 6

Exercise 6a – p. 91

1. 0.5317

2. 8.777

3. 0.041 91

4. 1.588

5. 30.8°

6. 33.7°

7. 51.3°

8. 40.9°

9. (a) 56.3°   (b) 68.2°   (c) 29.2°

10. 35.0°

11. 49.4°

12. 33.7°

13. 32.7°

14. 59.0°

15. 60.1°

16. 6.25

17. 1.75

18. 0.593

19. 3.30

20. 27.0

21. 18.2

Exercise 6b – p. 93

1. (a) 0.9703   (b) 0.3827
2. (a) 0.1045   (b) 0.9962
3. (a) 0.9455   (b) 0.3090   (c) 0.4540
   (d) 0.9781   (e) 0.1392   (f) 0.8660
4. (a) 27.8°   (b) 26.6°
   (c) 55.1°   (d) 48.2°
5. (a) 53.1°   (b) 28.1°   (c) 67.4°
6. (a) sine, 53.1°   (b) cosine, 45.6°
   (c) sine, 41.8°
7. 36.6°
8. 10.4 cm
9. 18.8 cm
10. 58.7°
11. 14.3 cm
12. 0.757 cm

Exercise 6c – p. 96

1. (a) 62.0°   (b) 33.0°   (c) 76.4°
   (d) 13.5°   (e) 31.8°, 7.97 cm   (f) 13.1 m
2. 12.1 m
3. 23.6°
4. 33.7°
5. 51.9 m
6. 157 km

Exercise 6d – p. 99

1. $\sin A = \frac{12}{13}$, $\cos A = \frac{5}{13}$
2. $\tan X = \frac{3}{4}$, $\sin X = \frac{3}{5}$
3. $\cos P = \frac{9}{41}$, $\tan P = \frac{40}{9}$
4. $\sin A = \dfrac{1}{\sqrt{2}} = \cos A$
5. $\sin Y = \frac{1}{3}\sqrt{5}$, $\tan Y = \frac{1}{2}\sqrt{5}$
6. $\cos A = \frac{1}{2}\sqrt{3}$; 30°
7. $\cos X = \frac{24}{25}$
8. $\cos X = \frac{4}{5}$
   $\cos^2 X + \sin^2 X = 1$
9. $\cos X = \frac{1}{2}\sqrt{3}$
   $\cos^2 X + \sin^2 X = 1$
10. $\cos X = \dfrac{1}{q}\sqrt{(q^2 - p^2)}$; 1
11. (a) $\frac{1}{2}(\sqrt{2} + \sqrt{3})$
    (b) $\frac{1}{2}(\sqrt{3} + 1)$
    (c) $2 + \sqrt{2}$
12. (a) 0.0707
    (b) 0.0966

13. 33.8 m
14. 570 m
15. $\dfrac{\sqrt{2} - 1}{\sqrt{4 - 2\sqrt{2}}}$; $\dfrac{1}{\sqrt{4 - 2\sqrt{2}}}$; $\sqrt{2} - 1$
16. 4.62 cm
17. 34.5 m

CHAPTER 7

Exercise 7a – p. 104

1.
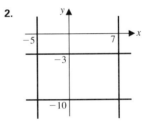

2.
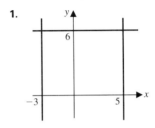

In Questions 3–10, the unshaded region is the one required.

3.
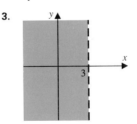

4.
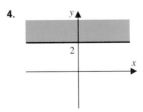

5.

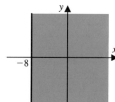

6.

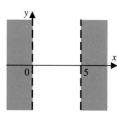

7.

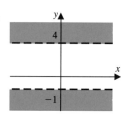

8.

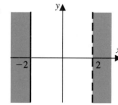

9.

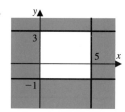

10.

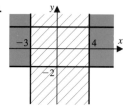

Exercise 7b – p. 107

1. (a) $y = 2x$ (b) $x + y = 0$
    (c) $3y = x$ (d) $4y + x = 0$
    (e) $y = 0$ (f) $x = 0$

2. (a) $2y = x + 2$ (b) $3y + 2x = 0$
    (c) $y = 4x$

3. (a) $2y = x$ (b) $y + 2x - 6 = 0$
    (c) $y = 2x + 5$

4. (a) $2y + x = 0$ (b) $2x - 3y = 0$
    (c) $2x + y = 0$

5. (a) $x - 3y - 1 = 0$ (b) $2x + y - 5 = 0$

6. (a) $5x - y - 17 = 0$
    (b) $x + 7y + 11 = 0$

7. $3x + 4y - 48 = 0$, 5

Exercise 7c – p. 111

1. (a) $y = 3x - 3$ (b) $5x + y - 6 = 0$
    (c) $x - 4y - 4 = 0$ (d) $y = 5$
    (e) $2x + 5y - 21 = 0$
    (f) $15x + 40y + 34 = 0$

2. (a) $3x - 2y + 2 = 0$
    (b) $3x - 2y + 7 = 0$
    (c) $x = 3$ (d) $2y = 5x$
    (e) $8x + 3y + 11 = 0$
    (f) $5x + 6y + 35 = 0$

3. (a), (c) and (d)

4. $x + y - 7 = 0$

5. $y = 3x + 5$

6. $x + y + 4 = 0$

7. $2x + y = 0$

8. $(0, 0), (0, 4), (-5, 0)$

9. $5x + 4y = 0$

10. $x + 2y - 11 = 0$

11. $3x - 4y + 19 = 0$

Mixed Exercise 7 – p. 113

1. (a) $\left(\frac{9}{8}, -\frac{13}{8}\right)$ (b) $\left(\frac{5}{7}, \frac{11}{7}\right)$
    (c) $\left(\frac{34}{5}, \frac{23}{5}\right)$

3. $\frac{1}{4}$ sq. units

4. $\left(-\frac{4}{3}, \frac{11}{3}\right), (-5, 0), (6, 0)$

5. $2x + y - 1 = 0$

6. $x + 3y - 11 = 0$; $\left(\frac{13}{5}, \frac{14}{5}\right)$

7. $y = x - 4$ (or $y + x = 10$ if line is inclined at $-45°$ to Ox)

8. $ay + 4bx - 5ab = 0$

9. (a) $\sqrt{20}$ (b) $x - 2y + 1 = 0$

10. $\left(\frac{9}{10}, \frac{17}{10}\right)$ and $\left(-\frac{18}{10}, \frac{26}{10}\right), \left(-\frac{27}{10}, -\frac{1}{10}\right)$
    or $\left(\frac{36}{10}, \frac{8}{10}\right), \left(\frac{27}{10}, -\frac{19}{10}\right)$

11. (a) $(1, 2), (5, 2), (3, 6)$ (b) 8

12. $\left[\frac{2}{5}(2 + 2a - b), \frac{1}{5}(4 - 2a + b)\right]$

13. $y = 2x - 3$
14. (a) $y = x$ (b) $\left(\frac{40}{13}, \frac{40}{13}\right)$
15. $y = \sqrt{3}x + 1 - 5\sqrt{3}$
16. (a) because the points lie (approximately) on a straight line
  (b) $a = 0.006$, $b = 0.2$
  (c) a is the length of the spring before a weight is attached
  (d) We cannot be sure that the spring continues to behave in the same way when heavier weights are attached.

CONSOLIDATION B

Multiple Choice Exercise B – p. 118
1. B
2. E
3. D
4. C
5. D
6. B
7. A
8. C
9. B
10. F
11. T
12. F
13. T
14. T
15. T

Miscellaneous Exercise B – p. 119
1. $3x + 5y - 4 = 0$, $(4/3, 0)$
2. (a) $y = (1/2)x + 3$ (b) $3\sqrt{5}$
3. $y = 3x - 1$, $(2/7, -1/7)$
4. (a) $2y = 3x - 9$ (b) $x + y = 3$
5. 31.4 km, 223.2°
6. $2y = x + 10$, $y = 3x - 5$, $(5, 5/2)$, $(2, 6)$
8. (a) AB $= \sqrt{32}$, BC $= \sqrt{72}$, AC $= \sqrt{104}$, 90°   (b) AB:1, BC:-1
9. $2y + 3x + 11 = 0$, $2y = x + 9$, $2y = 3x - 5$
10. (a) A$(-3, 0)$, B$(0, 3/2)$
  (b) $2y = x + 3$ (c) $(9/5, 12/5)$
11. 44.5° to 45.6°
12. 8.83 m, 2.07 m
13. 7.1°, 19.4°, 153.4°
14. $p = -7/2$, $q = -4$
16. (a) $5y = 2x$ (b) $(10, 4)$
17. (a) $(3, 3)$
18. $(4\sqrt{2}, 8\sqrt{2} + 4)$, $(12 + 4\sqrt{2}, 8\sqrt{2})$, $(0, 4)$
20. (a) 632.7, 659.7   (b) 2.1%

CHAPTER 8

Exercise 8a – p. 125
1. (a) ∠ADE, ∠ACE   (b) ∠AFE
  (c) ∠ADC   (d) ∠ABC   (e) ∠CDE
  (f) ∠CAD, ∠CED   (g) ∠BAC, ...
3. (b) 106.3°   (c) (5, 8), 53.1°

Exercise 8b – p. 128
1. (a) 50°   (b) 40°, 40°   (c) 60°
  (d) 54°, 108°
3. (4, 4)
4. 30°
7. $a^2 + b^2 - 9a - b + 14 = 0$
9. 60°, 60°

Exercise 8c – p. 132
1. 61.9°
2. 5 units
3. 67.4°
4. 7.22 cm
5. 60°, 30°, 90°
6. 8.43
7. ∠C $= 90°$, ∠A $= 18.4°$, ∠B $= 71.6°$
8. $\dfrac{2\sqrt{5}}{2}$
9. $\left(-\frac{3}{4}, -\frac{9}{4}\right)$
10. (a) $2x + y = 0$ (b) $\left(-\frac{4}{5}, \frac{8}{5}\right)$
11. $x + y - 14 = 0$
13. 16 sq. units
14. 14.9 cm

CHAPTER 9

Exercise 9a – p. 136
1. $\frac{1}{4}\pi$, $\frac{5}{6}\pi$, $\frac{1}{6}\pi$, $\frac{3}{2}\pi$, $\frac{5}{4}\pi$, $\frac{1}{8}\pi$, $\frac{4}{3}\pi$, $\frac{5}{3}\pi$, $\frac{7}{4}\pi$
2. 30°, 180°, 18°, 45°, 150°, 15°, 22.5°, 240°, 20°, 270°, 80°
3. (Angles in radians, correct to 3 sf) 0.611, 0.824, 1.62, 4.07, 0.246, 2.04, 6.46
4. (Angles to the nearest degree.) 97°, 190°, 57°, 120°, 286°, 360°
5. (a) 0.1987   (b) 0.4536   (c) 14.10
  (d) 1   (e) 0.7071   (f) 0.8660

Exercise 9b – p. 139
1. 2.09 cm, 4.19 cm^2
2. 26.2 cm, 131 cm^2
3. 4.77 cm, 35.8 cm^2
4. 7.96 cm, 79.6 cm^2
5. 18.8 cm, 75.4 cm^2

6. 3.14 cm, 1.57 rad
7. 4.8 cm, 0.96 rad
8. π cm, 6 rad
9. $\frac{5}{8}\pi$, 8 cm
10. 146°
11. 0.283 rad
12. 0.52°
13. (a) 12 cm^2 (b) 23.2 cm^2
14. 14.5 mm^2, 139 mm^2
15. (a) 15.2 cm   (b) 32.5 cm^2
16. 19.1°
17. 85.6 cm
18. 19.6 cm, 108 cm^2
20. 0.979a^2
21. (a) 135°   (b) 9.41 cm
    (c) 35.4 cm   (d) 68.8 cm

CHAPTER 10

Exercise 10a – p. 144
1. (a) yes   (b) yes   (c) yes, $x \neq 1$
    (d) no   (e) yes, $x \geqslant 0$ (f) yes
    (g) yes   (h) no
2. -4, -24
3. 25, 217
4. 1, not defined, 12
5. 1, $\dfrac{\sqrt{3}}{2}$

Exercise 10b – p. 147
1. (a) $f(x) \geqslant -3$ (b) $f(x) \geqslant -5$
    (c) $f(x) \geqslant 0$ (d) $0 < f(x) \leqslant \frac{1}{2}$
2. (a)

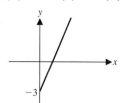

    (b)

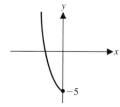

(c)

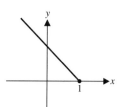

(d)

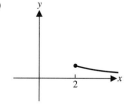

3. (a) 5, 4, 2, 0
    (b)

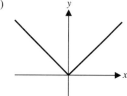

4. (a) 0, 2, 4, 5, 5
    (b)

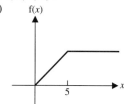

    (c) $0 \leqslant f(x) \leqslant 5$

5. (a) 0, £5000
    (b) $f(x) = 0$ for $0 \leqslant x \leqslant 20\,000$
        $f(x) = \frac{1}{5}(x - 20\,000)$ for $x > 20\,000$

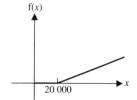

        domain $x \geqslant 0$ (but $x < $ GNP!)
        range $f(x) \geqslant 0$

Answers

Exercise 10c – p. 151

1. (a) $\frac{11}{4}$ (b) 3 (c) 4

2. (a) $f(x) \leqslant \frac{29}{4}$ (b) $f(x) \geqslant -2$
(c) $f(x) \leqslant 1$

3. (a)

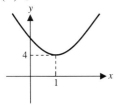

(b)

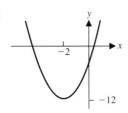

(c)

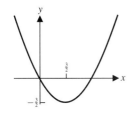

(d)

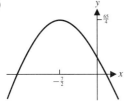

(e)

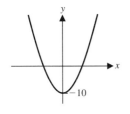

(f)

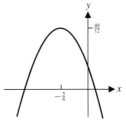

4. (a)

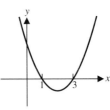

(b)

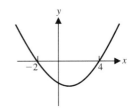

(c)

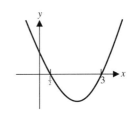

(d)

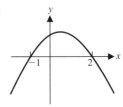

(e)

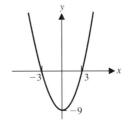

(f)

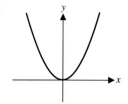

1.

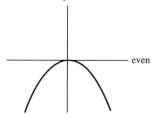

even

Exercise 10d – p. 153
1. (a)

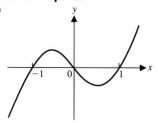

2.

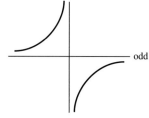

odd

(b)

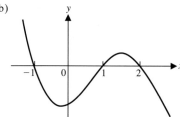

3.

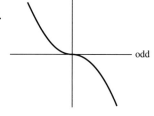

odd

2.

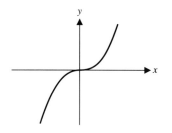

4.

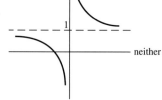

neither

5.

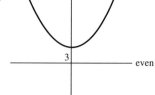

Exercise 10e – p. 153

1. (b) translation by c units in direction Oy
2. translation by c units in direction xO.
3. (a) reflection Ox
    (b) reflection Oy

even

6.

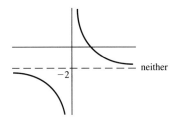

neither

7.

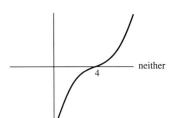

neither

8.

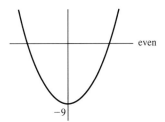

even

9.

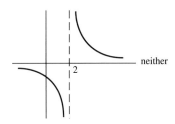

neither

10.

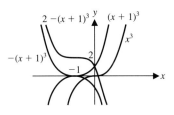

11. Reflection in line $y = x$
(there are several alternatives).
12. Reflection in line $y = x$
(there are several alternatives).
13. $(5, 2)$
14. (b, a)

Exercise 10g – p. 164
1. (a)

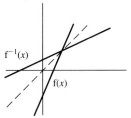

(b)

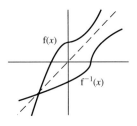

(c)

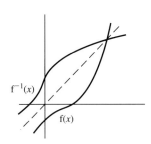

(d)

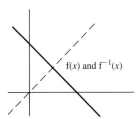

(e)

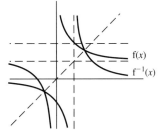

(f)

f(x) and f^{-1}(x)

3.

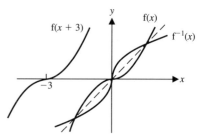

f(x + 3)

y

f(x)

f^{-1}(x)

−3

x

2. (d) and (f)

3. (a) f^{-1} = (x − 1)   (b) no
(c) f^{-1}(x) = $\sqrt[3]{(x-1)}$
(d) f^{-1}(x) = $\sqrt{(x+4)}$, x ⩾ −4
(e) f^{-1}(x) = $\sqrt[4]{x}$ − 1, x ⩾ 0

4. (a) −$\frac{1}{3}$ (b) $\frac{1}{2}$ (c) there isn't one

5. (a) 8   (b) 3   (c) 1

4. (a) g is even, h is odd, f is neither
(b) 9, $\frac{1}{9}$, 1
(c) $\dfrac{1}{2x^2+1} \cdot \left(\dfrac{x+2}{x}\right)^2$
(d) f^{-1}(x) = $\frac{1}{2}$(x − 1)
    g^{-1} does not exist
    h^{-1}(x) = $\dfrac{1}{x}$
(e) ±$\frac{1}{3}$
(f) No

Exercise 10h – p. 165

1. (a) $\dfrac{1}{x^2}$ (b) (1 − x)^2 (c) 1 − $\dfrac{1}{x}$

(d) 1 − x^2 (e) $\dfrac{1}{x^2}$

2. (a) 125   (b) 15   (c) −1   (d) −1

3. (a) (1 + x)^2 (b) 2(1 + x)^2
(c) 1 + 4x^2

4. g(x) = x^2, h(x) = 2 − x

5. g(x) = x^4, h(x) = (x + 1)

6. (a) f(x) = gh(x), g(x) = $\dfrac{1}{x}$,
    h(x) = 2x + 1
(b) f(x) = gh(x), g(x) = x^4,
    h(x) = 5x − 6
(c) f(x) = gh(x), g(x) = $\sqrt{x}$,
    h(x) = x^2 − 2

5. (a)

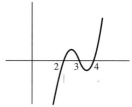

2 3 4

(b)

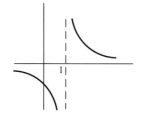

3

Mixed Exercise 10 – p. 166

1. (a) $\frac{1}{0}$ is meaningless   (b) $\frac{1}{4}$
(c)

1

(c)

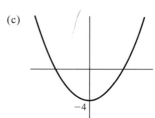

−4

(d) f^{-1}(x) = 1 − $\dfrac{1}{x}$, x ≠ 0

2. (a) $\frac{11}{4}$ when x = $\frac{3}{2}$
(b) −$\frac{41}{8}$ when x = $\frac{7}{4}$
(c) −9 when x = −2

6. (a) g(x) = $\dfrac{1}{x}$, h(x) = 2x − 1
(b) −3, $\frac{3}{4}$

7. (a)

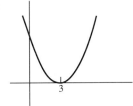

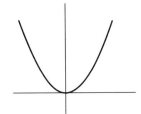

(b) $0, \frac{1}{6}$

8. (a)

(i) $\mathrm{f}(x) = \begin{cases} 25x \text{ for } 0 \leqslant x < 0.2 \\ 25 - 100x \text{ for } 0.2 < x < 0.3 \end{cases}$

(ii) $\mathrm{g}(x) = \begin{cases} 5 \text{ for } 0 \leqslant x < 0.2 \\ -5 \text{ for } 0.2 < x < 0.3 \end{cases}$

(b) (i)

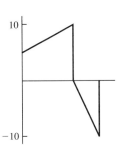

(ii)

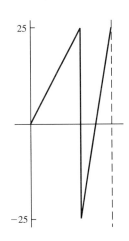

(iii)

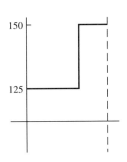

CHAPTER 11

Exercise 11a – p. 169

1. $x < \frac{7}{2}$

2. $x > 4$

3. $x > -3$

4. $x > -2$

5. $x > \frac{1}{2}$

6. $x < -3$

7. $x < -\frac{1}{4}$

8. $x > \frac{8}{3}$

9. $x > \frac{3}{8}$

Exercise 11b – p. 170

1. $1 < x < 3$

2. $-4 < x < -\frac{2}{3}$

3. $-1 < x < 2$

4. $-3 < x < 2$

5. $x < -13, x > -2$

6. $2 < x < 5$

7. $-1 < x < 4$

8. $4 < x < 5$

Exercise 11c – p. 172

1. $x > 2$ and $x < 1$

2. $x \geqslant 2$ and $x \leqslant -3$

3. $-4 < x < 2$

4. $x \geqslant \frac{1}{2}$ and $x \leqslant -1$

5. $x > 2 + \sqrt{7}$ and $x < 2 - \sqrt{7}$

6. $-\frac{1}{2} < x < \frac{1}{2}$

7. $-4 \leqslant x \leqslant 2$

8. $x > 1$ and $x < -\frac{2}{5}$

9. $x \geqslant \frac{3}{2}$ and $x \leqslant -5$

10. $x > 4$ and $x < -2$

11. $\frac{1}{2}(-3 - \sqrt{17}) \leqslant x \leqslant \frac{1}{2}(-3 + \sqrt{17})$

12. $x > 7$ and $x < -1$

Exercise 11d – p. 173

1. (a) $-1 < k < 1$

    (b) $k < 0$, $k > \frac{4}{9}$

2. (a) $p \geqslant 9$, $p \leqslant 1$

    (b) $p \geqslant 5$, $p \leqslant 1$

3. $-2 < a < 6$

4. (a) $p < 1$, $p > 9$

    (b) $p = 1$ and 9

    (c) $1 < p < 9$

Mixed Exercise 11 – p. 174

1. $x < 1$

2. $x > \frac{3}{2}$

3. $x > \frac{3}{2}$

4. $x > 3$, or $x < -2$

5. $-\frac{2}{3} < x < \frac{3}{2}$

6. $-\sqrt{13} < x < \sqrt{13}$

7. $x > 3 + \sqrt{2}$, or $x < 3 - \sqrt{2}$

8. $-2 < x < 7$

9. $1 < x < 4$

10. $-6 < x < -3$

11. $-1 < x < 2$

12. $-8 < x < -3$

14. $k \leqslant 3$, $k \geqslant 4$

15. (a)

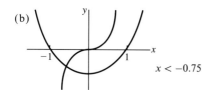

$0 < x < 1$, $x > 2$

(b)

$x < -0.75$

(c) From 1 to 4 days

CHAPTER 12

Exercise 12a – p. 176

1.

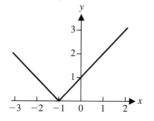

2.

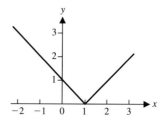

3.

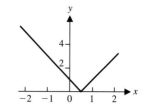

4.

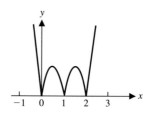

5.

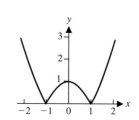

6.

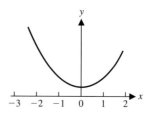

7.

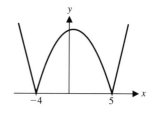

8.

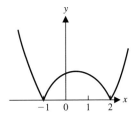

9.

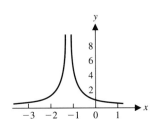

10.

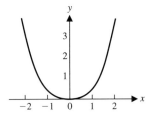

Exercise 12b – p. 178

1. $(-1, 1), (0, 0), (1, 1)$

2. $\left[\frac{1}{2}(\sqrt{5}-1), \frac{1}{2}(\sqrt{5}-1)\right],$
$\left[\frac{1}{2}(\sqrt{5}+1), \frac{1}{2}(\sqrt{5}+1)\right]$

3. $(1, 1), (-1, 1)$

4. $(2, 5), \left[\frac{1}{2}(3-\sqrt{41}), \frac{1}{2}(-7+3\sqrt{41})\right]$

5. $(1, 3), (1+\sqrt{6}, 3+2\sqrt{6})$

6. $\left(\frac{1}{6}\{-1-\sqrt{13}\}, \frac{1}{2}\{5+\sqrt{13}\}\right),$
$\left(\frac{1}{2}\{\sqrt{13}-1\}, \frac{1}{2}\{5-\sqrt{13}\}\right)$

7. $x = -\frac{1}{4}$

8. $x = 0$

9. $x = \frac{1}{2}$

10. $x = \frac{1}{2}(\sqrt{17}-3), x = 3$

11. $x = -1, x = -1-\sqrt{2}$

12. $x = 0, x = -\frac{1}{2}(5+\sqrt{57})$

Exercise 12c – p. 181

1.

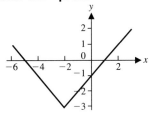

2.

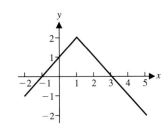

3.

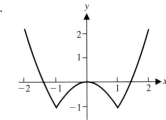

4.

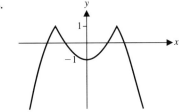

5. $-0.5 < x < 0.5$

6. $x > -0.5$

7. $-1 < x < 1/3$

8. $x > 1$

9. $x < 1$

10. $x < -1.25, x > -0.25$

11. $0 < x < 2.73$

12. $x < -1, x > 1$

13. (a)

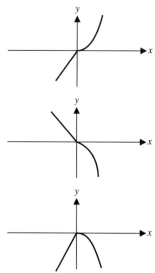

(b)

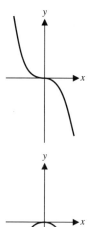

14. $|50 - 13t| < 20$ 3 hours 5 minutes
    (nearest minute)

CONSOLIDATION C
Multiple Choice Exercise C – p. 184
1. C
2. B
3. D
4. D
5. D
6. B
7. E
8. C
9. B
10. C
11. B
12. F
13. T
14. F
15. F
16. F
17. F
18. T
19. F
20. F
21. T

Miscellaneous Exercise C – p. 187
1. 24.6 m
2. $-4 < x < 4$
3. $x < -3$, $x > 4$
4. (a) $x \leqslant 3$ (b) $0 < k < 4$
   (c) f is one to one, g is not
   (d) $0 \leqslant x \leqslant 4$, $0 \leqslant f^{-1}(x) \leqslant 2$
5. (a) (i) $r\theta$ (ii) $\frac{1}{2}r^2\theta$
   (c) $r = 1$, $\theta = 2$ (d) E
6. (a)

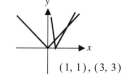

   (b) $f^{-1}(x) = \dfrac{8}{x} - 2$ (c) 2
7. $4.4 < x < 9$
8. $2(x + 5/4)^2 + 7/8$; 2, $\frac{5}{4}$, $\frac{7}{8}$; $(-5/4, 7/8)$
9. (a) $f(x) \geqslant -16$ (b) -2
   (c) $4 - \sqrt{(x + 16)}$
10. (a) $(10x + 1)/(x + 5)$, $x \neq -5$
    (b) $f^{-1}(x) = (2x + 1)/(x - 3)$, $x \neq 3$
11.

$(1, 1)$, $(3, 3)$

12.

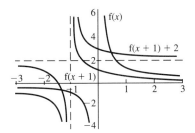

13. (a) (1) and (3)   (b) (2) and (6)
      (c) (1) and (4)
14.

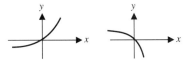

15. (a) $g(x) \geqslant 1$ (b) 26   (c) 10, 1
      (d) 3, $-2\frac{1}{3}$
16. (a) $\pm 2, \pm 4$
      (b)   :3, -3

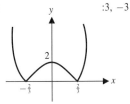

17. (a) $x < 1/2$
18. $x < -1.30, x > 2.30$

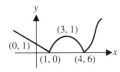

19. (a)

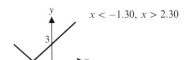

      (b)

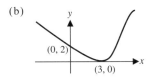

(c)

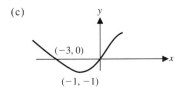

(d)

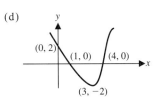

19. 0.6
20. (a) e.g. $x^4 + 1$
      (b) (i) two turning points: looks like a
               cubic
          (ii) $(0, -4.5), (1, 0), (3, 0)$;
               $p = 1, q = 3, a = 0.5$
          (iii) $t = 2$: model $s = 1/2$,
               graph $s \simeq 0.8$
               $t = 4$: model $s = 1.5$,
               graph $s \simeq 1$
               $t = 5$: model $s = 7$,
               graph $s \simeq 4$
               Not good, very poor for $t > 4$
          (iv) $1/10$
22. $0.725 \, \text{cm}^2$

CHAPTER 13

Exercise 13a – p. 201
  1. $5x^4$
  2. $11x^{10}$
  3. $20x^{19}$
  4. $10x^9$
  5. $-3x^{-4}$
  6. $-7x^{-8}$
  7. $-5x^{-6}$
  8. $\frac{4}{3}x^{1/3}$
  9. 1
 10. $-x^{-2}$
 11. $\frac{1}{3}x^{-2/3}$
 12. $\frac{1}{2}x^{1/2}$ or $\frac{1}{2\sqrt{x}}$
 13. $\frac{-2}{x^3}$
 14. $\frac{-1}{x^{3/2}}$
 15. $\frac{-4}{x^5}$

16. $\dfrac{-10}{x^{11}}$

17. $-\frac{1}{4}x^{-5/4}$

18. $\frac{5}{2}\sqrt{x^3}$

19. $\frac{1}{7}x^{-6/7}$

20. px^{p-1}

Exercise 13b – p. 203

1. $15x^2$

2. 7

3. $-\dfrac{8}{x^2}$

4. $\dfrac{5}{2\sqrt{x}}$

5. $\dfrac{3}{2x^4}$

6. $9x^2 - 8x$

7. $4 + \dfrac{4}{x^2}$

8. $-\dfrac{3}{x^2} + \dfrac{1}{3}$

9. $3x^2 - 2x + 5$

10. $6x + \dfrac{4}{x^2}$

11. $3x^2 - 4x - 8$

12. $8x^3 - 8x$

13. $2x + \dfrac{5}{2\sqrt{x}}$

14. $9x^2 - 8x + 9$

15. $\frac{3}{2}x^{1/2} - \frac{1}{2}x^{-1/2} - \frac{1}{2}x^{-3/2}$

16. $\dfrac{1}{2\sqrt{x}} + \dfrac{3\sqrt{x}}{2}$

17. $-\dfrac{2}{x^3} + \dfrac{3}{x^4}$

18. $\dfrac{-1}{2\sqrt{x^3}} + \dfrac{2}{x^2}$

19. $-\frac{1}{2}x^{-3/2} + \frac{9}{2}x^{1/2}$

20. $\frac{1}{4}x^{-3/4} - \frac{1}{5}x^{-4/5}$

21. $-\dfrac{12}{x^4} + \dfrac{3x^2}{4}$

22. $-\dfrac{4}{x^2} - \dfrac{10}{x^3} + \dfrac{18}{x^4}$

23. $\dfrac{3}{2\sqrt{x}} - 3$

24. $1 + 2x^{-2} + 9x^{-4}$

Exercise 13c – p. 204

1. $\dfrac{dy}{dx} = 2x + 2$

2. $\dfrac{dz}{dx} = -4x^{-3} + x^{-2}$

3. $\dfrac{dy}{dx} = 6x + 11$

4. $\dfrac{dy}{dz} = 2z - 8$

5. $\dfrac{ds}{dt} = -\dfrac{3}{2t^4}$

6. $\dfrac{ds}{dt} = \dfrac{1}{2}$

7. $\dfrac{dy}{dx} = 1 - \dfrac{1}{x^2}$

8. $\dfrac{dy}{dz} = \dfrac{5z^2 - 1}{2\sqrt{z}}$

9. $\dfrac{dy}{dx} = 18x^2 - 8$

10. $\dfrac{ds}{dt} = 2t$

11. $\dfrac{ds}{dt} = 1 - \dfrac{7}{t^2}$

12. $\dfrac{dy}{dx} = -\dfrac{3\sqrt{x} + 28}{2x^3}$

Exercise 13d – p. 206

1. 2

2. $-\frac{1}{3}$

3. $\frac{1}{4}$

4. 6

5. 1

6. 5

7. 11

8. -11

9. 4

10. $\frac{4}{27}$

11. $\frac{5}{4}$

12. 2

13. $(2, 2)$ and $(-2, 4)$

14. $(1, 0)$ and $\left(-\frac{1}{3}, \frac{4}{27}\right)$

15. $(3, 0)$ and $(-3, 18)$

16. $(-1, -2)$ and $(1, 2)$

17. $(1, -16)$

18. $\left(-1, \frac{1}{4}\right)$

19. $(0, -5)$

20. $(1, -2)$ and $(-1, 2)$

Mixed Exercise 13 – p. 207

1. (a) $3x^2$ (b) $4x$ (c) $-\dfrac{1}{2x^2}$

    (d) $-\dfrac{1}{x^{3/2}}$ (e) $-\dfrac{1}{2x^3}$ (f) $-\dfrac{2}{\sqrt{x}}$

2. (a) $6x+1$ (b) $3x^2-3$ (c) $2x+1$

3. (a) $-3x^{-4}-3x^2$

    (b) $\frac{1}{2}x^{-1/2}+\frac{1}{2}x^{-3/2}$

    (c) $-\dfrac{2}{x^3}-\dfrac{6}{x^4}$

4. (a) $\frac{3}{2}x^{1/2}-\frac{2}{3}x^{-1/3}-\frac{1}{3}x^{-4/3}$

    (b) $\dfrac{1}{2\sqrt{x}}+\dfrac{1}{x^2}-\dfrac{3}{x^4}$

    (c) $\dfrac{1}{x^2}-\dfrac{1}{2x^{3/2}}$

    (d) $-\dfrac{1}{2\sqrt{x}}-\dfrac{3\sqrt{x}}{2}$

5. (a) 5   (b) 5   (c) 17

6. (a) 5   (b) 6   (c) 12   (d) $-\frac{5}{4}$

7. (a) 1   (b) 7 and -7

8. (a) -3 (b) $\left(-\frac{1}{2},0\right)$ and $(2,0)$

    (c) -5 and 5

9. (a) $(1,10)$ and $(-1,6)$

    (b) $\left(\frac{1}{3},7\frac{7}{9}\right)$ and $\left(-\frac{1}{3},8\frac{2}{9}\right)$

10. (a) $4x^3-2x$ (b) $6x+8$

    (c) $\dfrac{x+3}{2x\sqrt{x}}$

11. -2

12. $\left(\frac{1}{8},\frac{1}{64}\right)$

13. 5 and -5

14. $(1,9)$ and $(3,11)$

16. (b), (d)

17. 11, $y=11x-6$; $\left(\frac{2}{3},\frac{4}{3}\right)$

18. $2x+y-2=0$

19. $(2,14),(-2,-14)$

20. $(1,9),(3,11)$

CHAPTER 14

Exercise 14a – p. 210

1. $x=0$

2. $x=\frac{3}{4}$

3. $x=0$, $x=\frac{8}{3}$

4. $x=\frac{3}{2}$

5. $x=0$, $x=\frac{4}{3}$

6. $x=\pm1$

7. $x=4$

8. $x=\pm3$

9. $x=1$, $x=-\frac{4}{3}$

10. $x=\pm\frac{5}{9}\sqrt{3}$

11. $x=2$

12. $x=\pm\frac{2}{3}\sqrt{3}$

13. $(3,3),(-3,-3)$

14. $(1,-7),\left(\frac{1}{3},-\frac{185}{27}\right)$

15. $\left(\frac{1}{2},-\frac{25}{4}\right)$

16. $(4,0)$

17. $(1,2)$

18. $(4,10),(-4,6)$

Exercise 14b – p. 214

1. $(1,1)$

2. $(-1,-2)$ min; $(1,2)$ max

3. $(3,6)$ min; $(-3,-6)$ max

4. $(0,0)$ max; $\left(\frac{10}{3},-\frac{500}{27}\right)$ min

5. $(0,0)$ min

6. $\left(\frac{5}{2},0\right)$ min

7. $(0,0)$ inflex

8. $(0,0)$ min

9. $\left(\frac{5}{4},-\frac{49}{8}\right)$ min

10. $(-1,4)$ max; $(1,-4)$ min

11. -2 max; 2 min

12. $2\frac{3}{4}$ min

13. 54 max; -54 min

14. $-\frac{5}{16}$ min; 0 max; -2 min

15.

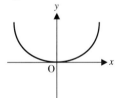

16. (a)

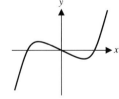

    (b)

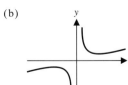

Exercise 14c – p. 218

1. $800\,\text{m}^2$; $20\,\text{m} \times 40\,\text{m}$

2. $20\,\text{cm} \times 20\,\text{cm} \times 10\,\text{cm}$

3. $r = \sqrt{(9 - h^2)}$; $12\pi\sqrt{3}\,\text{cm}^2$

4. $5\,\text{cm}$ square

5. $\sqrt{35}\,\text{cm}$ square

6. $a = 1$, $b = -2$, $c = 3$

7. $p = q = 1$, $r = 2$; $(-1,\,0)$

8. $5y = x^2 + 4x + 9$

9. $(-2,\,16)$ max, $(2,\,-16)$ min;

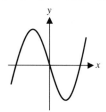

10. min 2, max -2;

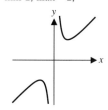

12. $h = \frac{1}{2}(7 - 2r - \pi r)$

13. $4\,\text{m} \times 4\,\text{m} \times 2\,\text{m}$

14. $12.5\,\text{cm}^2$

15. (a) (i) 600–1000 m below surface and 200–400 m below surface
    (ii) 400–600 m below surface
 (b) Seeping back into rock
 (c)

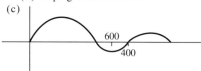

 (d) From surface to about 600 m below surface.

16.

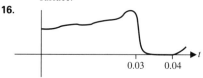

Yes, about 1 to $1\frac{1}{2}$ miles through the roadworks,
speed $\simeq \frac{0.5}{0.005}$ m.p.h. ($\simeq 100\,\text{m.p.h.}$)

17. (a)

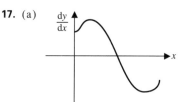

(b)

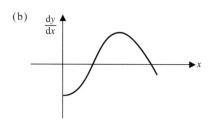

(c)

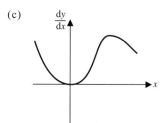

(d)

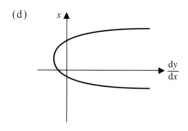

CHAPTER 15

Exercise 15a – p. 225

1. $\frac{1}{2}\sqrt{3}$

2. 0

3. $-\frac{3}{2}\sqrt{3}$

4. $\frac{1}{2}$

5. $-\frac{1}{2}\sqrt{3}$

6. $\frac{1}{2}$

7. $-1/\sqrt{2}$

8. $-\frac{1}{2}\sqrt{3}$

9. $\frac{1}{2}\pi$, $\frac{5}{2}\pi$, $\frac{9}{2}\pi$

10. $-\frac{1}{2}\pi$, $-\frac{5}{2}\pi$

11. sin 55°
12. −sin 70°
13. −sin 60°
14. −sin $\frac{1}{6}\pi$
15.

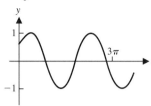

16.

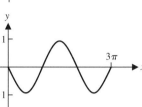

17.

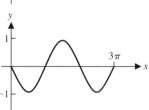

18.

19.

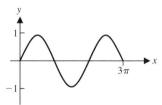

20.

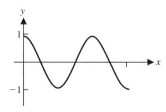

21. The curve $y = a \sin \theta$ is a one-way stretch of the curve $y = \sin \theta$ by a factor a parallel to the y-axis.
22. The curve $y = \sin 3\theta$ is a one-way shrinkage of the curve $y = \sin \theta$ by a factor $\frac{1}{3}$ parallel to the x-axis.
24. (a)

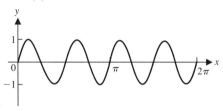

(b)

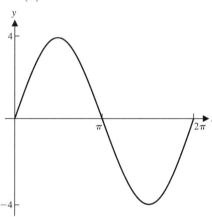

Exercise 15b – p. 229

1. (a) −cos 57°   (b) −cos 70°
     (c) cos 20°   (d) −cos 26°

2. (a) $-\dfrac{\sqrt{3}}{2}$ (b) 0   (c) $-\dfrac{1}{\sqrt{2}}$

     (d) 1   (e) $-\frac{1}{2}$ (f) $\frac{1}{2}$

     (g) $\dfrac{-3}{\sqrt{2}}$ (h) $\dfrac{-1}{\sqrt{2}}$

3. (a)

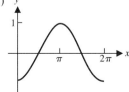

(b)

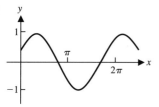

(c)

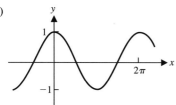

4.

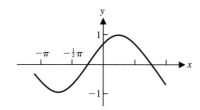

(a) $\theta = \frac{1}{4}\pi$ (b) $\theta = -\frac{3}{4}\pi$

(c) $\theta = -\frac{1}{4}\pi$ and $\frac{3}{4}\pi$

5.

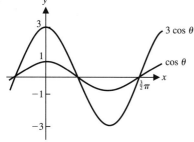

6.

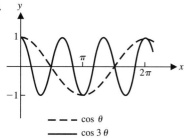

7.

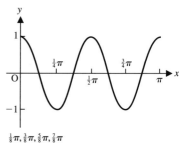

$\frac{1}{8}\pi, \frac{3}{8}\pi, \frac{5}{8}\pi, \frac{7}{8}\pi$

8.

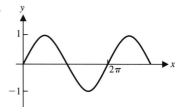

$\sin\theta = \cos(\theta - \frac{1}{2}\pi)$
$\cos\theta = -\sin(\theta - \frac{1}{2}\pi)$

9. (a) $\frac{1}{4}\pi, \frac{3}{4}\pi, \frac{5}{4}\pi, \frac{7}{4}\pi$
   (b) $\frac{1}{2}\pi, \frac{3}{2}\pi$

10. $\theta \simeq \frac{5}{8}\pi$ ($\pi = 1.95\,\text{rad}$)

11. 3, 5

Exercise 15c – p. 231

1. (a) 1   (b) $-\sqrt{3}$ (c) $\sqrt{3}$ (d) $\sqrt{3}$

2. (a) $\tan 40°$ (b) $-\tan \frac{2}{7}\pi$
   (c) $-\tan 50°$ (d) $\tan \frac{2}{5}\pi$

3. (a) $\frac{1}{4}\pi, \frac{5}{4}\pi$ (b) $\frac{3}{4}\pi, \frac{7}{4}\pi$
   (c) $0, \pi, 2\pi$ (d) $\frac{1}{2}\pi, \frac{3}{2}\pi$

Exercise 15d – p. 234

1. (a) 0.412 rad, 2.73 rad, −5.87 rad,
          −3.55 rad
   (b) $-\frac{4}{3}\pi, -\frac{2}{3}\pi, \frac{2}{3}\pi, \frac{4}{3}\pi$
   (c) 0.876 rad, 4.02 rad, −2.27 rad,
          −5.41 rad

2. (a) 141.3°, 321.3°, 501.3°, 681.3°
   (b) 191.5°, 348.5°, 551.5°, 708.5°
   (c) 84.3°, 275.7°, 444.3°, 635.7°

3. (a) 36.9°   (b) 323.1°   (c) 0.464 rad

4. 0, π, 2π

5. 11.8°, 78.2°, 191.8°, 258.2°

6. $\frac{1}{3}\pi, \pi, \frac{5}{3}\pi$

7. (a)

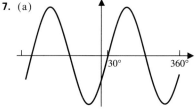

(b)

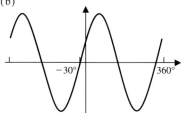

8. (a)

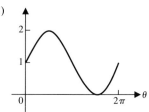

(b)

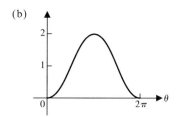

(c)

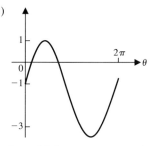

9. (a) 3 (b) 2, or any even number

Exercise 15e – p. 236
1. $22.5°, 112.5°, 202.5°, 292.5°$
2. $40°, 80°, 160°, 200°, 280°, 320°$
3. no values in the given range

4. $12°, 60°, 84°, 132°, 156°, 204°, 228°, 276°,$
    $300°, 348°$
5. no values in the given range
6. $25.5°, 154.5°, 205.5°, 334.5°$
7. $135°, 315°$
8. $240°$
9. $60°, 240°$
10. $0, 120°, 360°$
11. $-\frac{7}{12}\pi, \frac{1}{12}\pi$
12. $-\frac{11}{12}\pi, \frac{1}{12}\pi$
13. $0, \frac{2}{3}\pi$
14. $-\frac{1}{3}\pi, \pi$
15. $149.5°, 59.5°, 30.5°, 120.5°$
16. $-105.2°, 14.8°, 134.8°, -74.8°, 45.2°,$
    $165.2°$
17. $\pm 63.6°$
18. $\frac{1}{6}\pi, \frac{5}{12}\pi, \frac{2}{3}\pi, \frac{11}{12}\pi, \frac{7}{6}\pi, \frac{17}{12}\pi, \frac{5}{3}\pi, \frac{23}{24}\pi$
19. $\frac{1}{15}\pi, \frac{1}{3}\pi, \frac{7}{15}\pi, \frac{11}{15}\pi, \frac{13}{15}\pi, \frac{17}{15}\pi, \frac{19}{15}\pi, \frac{23}{15}\pi,$
    $\frac{5}{3}\pi, \frac{29}{15}\pi$
20. $\frac{3}{2}\pi$

CHAPTER 16
Exercise 16a – p. 238
1.   1.0299 rad

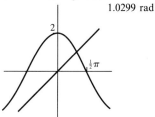

2.

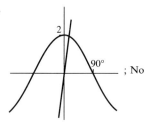

; No

3. (a) $-1.895\,49, 0, 1.895\,49$
   (b) $0, 0.8767$ rad   (c) 1.2834 rad

Exercise 16b – p. 239
1. $-0.804. \ 1.55$
2. $-2.24, 0, 2.24$
3. $0.338, 1.31$
4. $1.92, -1.58$

5. 1.29, 2.71
6. -2.56, 1, 1.56
7. 0, 1.57 rad
8. 0.666 rad, 2.48 rad
9. -3.14 rad, 0, 3.14 rad
10. -2.36 rad, 0.785 rad

Exercise 16c – p. 242

1. (a) 3   (b) 2   (c) 2   (d) infinite
    (e) 3   (f) 1
2. (a) $0.75 < x < 1.25$ (b) $1 < x < 1.5$
    (c) $0.5 < x < 1$ (e) $2 < x < 2.5$
    (f) $1 < x < 1.5$
3. $(0, 1)$ max, $(2, -3)$ min, 3
4. (a) 2   (b) 3   (c) 3
5. $-3 < x < -2$

CONSOLIDATION D

Multiple Choice Exercise D – p. 246

  1. C
  2. E
  3. B
  4. D
  5. A
  6. C
  7. B
  8. F
  9. T
10. T
11. T
12. F
13. F
14. T
15. F
16. T
17. T
18. F

Miscellaneous Exercise D – p. 248

1. (a) $3x^2 - 1$ (b) 6   (c) 2
    (d) $y = 2x + 8$ (e) $(1, 6)$
2. (a) -4 (b) 6   (c) $5\pi/4$, $7\pi/4$
3. $(2, 4)$ min, $(-2, -4)$ max; $x < -2$, $x > 2$
4. (a) $n - \alpha$ (b) $3\pi + \alpha$, $4\pi - \alpha$
5. $3x^2 - 8x + 5$ (a) 4   (b) 8
    (c) $y = 8x - 20$ (d) $8y + x - 35 = 0$
6. Least: -2, greatest: $1/4$
7. $x = 10/\sqrt{3}$, min $C = 20\sqrt{3}$, i.e. £35 000
    to nearest £1000
8. $1 < t < 4$
9. 60 km/h, £50

11. (a) $x = 1/3$, $V_{max} = 16/27$
    (b)

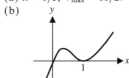

12. 20 m/s
13. (a) $\pi/6$, $5\pi/6$ (b) $\pi/12$, $13\pi/12$
15. $-1/3$, $1/2$; $60°$, $109.5°$, $250.5°$, $300°$
16. (a) There are two values of x for which
    y (i.e. $x^4 - 4x - 1) = 0$
    (b) -1, 1
17. (a) $-165°$, $-105°$, $15°$, $75°$
    (b) $-30°$, $90°$
18.

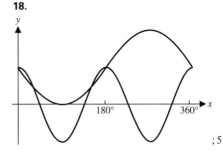

                                      ; 5

19. $(1.25, 1.5)$

CHAPTER 17

All general integrals in this chapter require
the addition of **K**, the constant of integration.

Exercise 17a – p. 253

  1. $\frac{1}{5}x^5$
  2. $\frac{1}{8}x^8$
  3. $\frac{1}{4}x^4$
  4. $\frac{1}{12}x^{12}$
  5. $-\dfrac{1}{x}$
  6. $\dfrac{-1}{4x^4}$
  7. $\frac{2}{3}x^{3/2}$
  8. $\frac{4}{7}x^{7/4}$
  9. $\frac{3}{4}x^{4/3}$
10. $-\frac{1}{2}x^{-2}$
11. $2x^{1/2}$
12. $\frac{1}{2}x^2$

Exercise 17b – p. 254

1. $\frac{1}{6}x^6 + \frac{2}{3}x^{3/2}$

2. $-\frac{1}{4x^4} - \frac{1}{3}x^3$

3. $\frac{4}{5}x^{5/4} + \frac{1}{5}x^5$

4. $-\frac{1}{2}x^{-2} - \frac{1}{4}x^4$

5. $\frac{-2}{3x^{3/2}} + \frac{5}{7}x^{7/5}$

6. $2x^{1/2} + \frac{2}{3}x^{3/2}$

7. $\frac{1}{2}x^2 + \frac{1}{x}$

8. $\frac{3}{2}x^{2/3} + \frac{2}{5}x^{5/2}$

Exercise 17c – p. 259

1. $\frac{1}{5}$

2. $\frac{1}{2}$

3. $\frac{14}{3}$

4. $\frac{15}{64}$

5. 2

6. $16\frac{1}{3}$

7. 4

8. $\frac{2}{7}(8\sqrt{2}, -1)$

9. $26\frac{2}{3}$

10. $17\frac{2}{3}$

11. 15

12. -2

13. $-\frac{9}{4}$

14. -2

15. $\frac{1}{3}$

16. $\dfrac{2(47\sqrt{2} - 28)}{15}$

Exercise 17d – p. 260

Answers in square units.

1. $5\frac{1}{3}$

2. $12\frac{2}{3}$

3. $2\frac{2}{3}$

4. $13\frac{1}{2}$

5. $5\frac{1}{3}$

6. 60

7. $5\frac{1}{3}$

8. $4\frac{7}{8}$

9. 24

Exercise 17e – p. 263

1. $2\frac{1}{3}$

2. $2\frac{1}{4}$

3. $28\frac{1}{2}$

4. $15\frac{1}{4}$

5. $2\frac{1}{6}$

6. $\frac{4}{3}$

7. $\frac{1}{3}$

8.

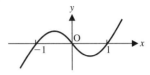

(a) $\frac{1}{4}$ (b) $\frac{1}{4}$ (c) $\frac{1}{2}$

9.

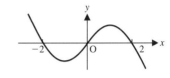

(a) 4 (b) 4 (c) 8

10.

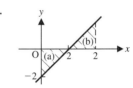

(a) -2 (b) 2 (c) 0

Exercise 17f – p. 264

1. $49\frac{1}{2}$

2. $2\frac{1}{3}$

3. $5\frac{1}{3}$

4. 36

5. $4\frac{1}{2}$

6. $A = \int_0^1 x\,\mathrm{d}y \qquad B = \int_0^1 y\,\mathrm{d}x$

   $A + B = 1$

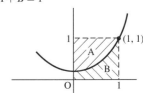

Mixed Exercise 17 – p. 265

1. $\frac{1}{3}x^3 + \frac{1}{x} + K$

2. $3x^3 + 21x^2 + 49x + K$

3. $\frac{2}{3}x^{3/2} + 2x^{1/2} + K = \frac{2}{3}\sqrt{x}(x+3) + K$

4. $\frac{2}{3}(2x-3)\sqrt{x} + K$

5. $\frac{1}{4}x^4 + \frac{1}{2x^2} + K$

6. $\frac{2}{5}\sqrt{x}(x^2 - 5) + K$

7. $\frac{1}{3}$

8. 12

9. $33\frac{3}{4}$

10. $20\frac{5}{6}$

11. 27

12. $\frac{3}{4}$

13. 16

14. $10\frac{2}{3}$

15. $\frac{1}{2}$

16. (b) $-57\frac{1}{6}$ (c) $57\frac{1}{6}$

17. (a)

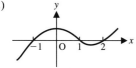

(b) $2\frac{1}{4}$

(c) $3\frac{1}{12}$

18. (a) $A_1 = \int_0^1 \frac{\sin x}{x}\, dx$ (b) $A_1 = y_1 - y_0$

(c)

x	0	1	2	3	4	5	6
y	0.9	1.84	2.51	2.75	2.66	2.45	2.32

(d)

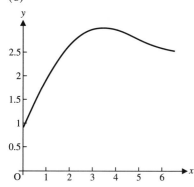

19.

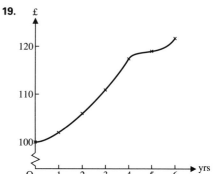

20.

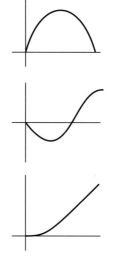

CHAPTER 18

Exercise 18a – p. 271

1. (a) $1, \frac{1}{2^2}, \frac{1}{3^3}, \frac{1}{4^4}, \frac{1}{5^5}, \frac{1}{6^6}$; converges
    (rapidly to 0)

(b) $-2, 4, -8, 16, -32, 64$; diverges

(c) $-\frac{1}{2}, \frac{1}{4}, -\frac{1}{8}, \frac{1}{16}, -\frac{1}{32}, \frac{1}{64}$; converges
    (to 0)

(d) $\frac{1}{2}, \frac{\sqrt{3}}{2}, 1, \frac{\sqrt{3}}{2}, \frac{1}{2}, 0$; periodic, cycles over
    12 terms

2. (a) $3, 0.57, 0.73, 0.70, 0.71, 0.71$; converges

(b) $-4, -10, -88, -7654, -5.9 \times 10^7$,
    -3.4×10^{15}; diverges (rapidly)

(c) $0.5, 1.2, 0.88, 1.1, 0.97, 1.0$; converges
    (to 1)

(d) $1, 0.8, 0.87, 0.85, 0.86, 0.85$; converges

3. (a) Undefined after u_2
(b) Cycles thro' $\frac{1}{2}$, $1\frac{1}{3}$, 3, -2
(c) Cycles thro' -2, $\frac{1}{2}$, $1\frac{1}{3}$, 3
4. (a) converges (to 0.4)
(b) converges (to 0.4)
(c) converges (to 0.4)
5. $\frac{1}{2}(3 \pm \sqrt{13})$

Exercise 18b – p. 274

1. (a) $\displaystyle\sum_{r=1}^{5} r^3$ (b) $\displaystyle\sum_{r=1}^{10} 2r$

(c) $\displaystyle\sum_{r=2}^{50} \frac{1}{r}$ (d) $\displaystyle\sum_{r=0}^{\infty} \frac{1}{3^r}$

(e) $\displaystyle\sum_{r=0}^{7} (-4 + 3r)$

(f) $\displaystyle\sum_{r=0}^{\infty} \left(\frac{8}{2^r}\right) = \sum_{r=0}^{\infty} \frac{1}{2^{r-3}}$

2. (a) $1 + \frac{1}{2} + \frac{1}{3} + \ldots$
(b) $0 + 2 + 6 + 12 + \ldots + 30$
(c) $2 + \frac{1}{2} + \frac{4}{15} + \ldots + \frac{22}{861}$
(d) $1 + \frac{1}{2} + \frac{1}{3} + \ldots$
(e) $0 + 0 + 6 + 24 + 60 + \ldots + 720$
(f) $-1 + a - a^2 + \ldots$
3. (a) 8; 9   (b) 9; 10   (c) -1; 6
(d) $\frac{1}{420}$; ∞ (e) $\left(\frac{1}{2}\right)^n$; ∞
(f) -48; 123   (g) 4; 10

Exercise 18c – p. 279

1. (a) 9, $2n - 1$ (b) 16. $4(n - 1)$
(c) 15, $3n$ (d) 17, $3n + 2$
(e) -2, $8 - 2n$ (f) $p + 4q, p + (n - 1)q$
(g) 18, $8 + 2n$ (h) 17, $4n - 3$
(i) 0, $\frac{1}{2}(5 - n)$ (j) 8, $3n - 7$
2. (a) 100   (b) 180   (c) 165
(d) 185   (e) -30 (f) $5(2p + 9q)$
(g) 190   (h) 190   (i) $-\frac{5}{2}$ (j) 95
3. $a = 27.2$, $d = -2.4$
4. $d = 3$; 30
5. 1, $\frac{1}{2}$, 0; $-8\frac{1}{2}$
6. (a) $28\frac{1}{2}$ (b) 80   (c) 400
(d) 80   (e) 108   (f) $3n(1 - 6n)$
(g) 40   (h) $2m(m + 3)$
7. 4, $2n - 4$
9. 2, 364
10. 39

11. 64
12. (a) 1, 5   (b) 270
13. (a) $a = 21$, $d = -3$
(b) less than 4 or more than 11

Exercise 18d – p. 284

1. (a) 32, 2^n (b) $\frac{1}{8}$, $\dfrac{1}{2^{n-2}}$
(c) 48, $3(-2)^{n-1}$
(d) $\frac{1}{2}$, $(-1)^{n-1}\left(\frac{1}{2}\right)^{n-4}$ (e) $\frac{1}{27}$, $\left(\frac{1}{3}\right)^{n-2}$
2. (a) 189   (b) -255 (c) $2 - \left(\frac{1}{2}\right)^{19}$
(d) $781/125$ (e) $341/1024$ (f) 1
3. $\frac{1}{2}$, 2
4. $-\frac{1}{2}$
5. $-\frac{1}{2}$, $1/1024$
6. 13.21 to 4 sf
7. (a) $(x - x^{n+1})/(1 - x)$
(b) $(x^n - 1)/x^{n-2}(x - 1)$
(c) $(1 + (-1)^{n+1}y^n)/(1 + y)$
(d) $x(2^n - x^n)/2^{n-1}(2 - x)$
(e) $[1 - (-2)^n x^n]/(1 + 2x)$
8. $\frac{8}{3}\left(1 - \left(\frac{1}{4}\right)^n\right)$, 4
9. 62 or 122
10. 8.493 to 4 s.f.
11. 8 (last repayment is less than £2000)
12. £23.31

Exercise 18e – p. 288

1. (a) yes   (b) no   (c) yes
(d) yes   (e) no   (f) yes
2. (a) 6   (c) $13\frac{1}{3}$ (d) $\frac{5}{9}$ (f) $\frac{9}{4}$
3. (a) $161/990$ (b) $34/99$ (c) $7/330$
4. $\frac{1}{2}$
5. 8, 4, 2, 1

Mixed Exercise 18 – p. 289

1. 1, $\frac{1}{4}$, $\frac{1}{9}$, $\frac{1}{16}$, $\frac{1}{25}$; converges (to zero)
2. $-\frac{1}{3}$, 0, $\frac{1}{5}$, $\frac{1}{3}$, $\frac{3}{7}$; converges to 1
3. 1, -1, 1, -1, 1; cycles between 1 and -1
4. -1, 1, -1, 1, -1; cycles between 1 and -1
5. 1, 3, 5, 7, 9; diverges
6. 1, 2, 4, 8, 16; diverges
7. 4, 2, $2^{1/2}$, $2^{1/4}$, $2^{1/8}$; converges to 1
8. 1, 2, 3, 5, 8; diverges
9. 1, 2, 2, 1, $\frac{1}{2}$; cycles through 1, 2, 2, 1, $\frac{1}{2}$, $\frac{1}{2}$
10. $\frac{2}{3}$

11. $\frac{1}{2}(1+3^{11}) = 88\,574$

12. $\dfrac{ab^4(1-b^{2(n-1)})}{1-b^2}$

13. $2(n+5)(n-4)$

14. $\dfrac{e(1-e^n)}{1-e}$

15. $\frac{1}{2}n(7+3n)$

16. $-3n^2$

17. 1

18. $n(n+1)$

19. $\frac{17}{330}$

20. $\frac{10}{99}$

21. 1, 7, 19, 37; $3n^2-3n+1$

22. 16, -8

23. 2

24. 1

25. $a = 2\pm\sqrt{2},\, r = \frac{1}{4}(2\pm\sqrt{2})$

26. £805 056.27

27. £6391.62

28. £725 000

29. About 28 years

CHAPTER 19

Exercise 19a – p. 292

1.

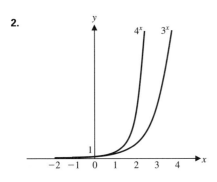

2.

3. (a) They are the same because
$2^{2x} = (2^2)^x = 4^x$
   (b) For all values of a, $y = a^x$ goes
   through $(1, 0)$ and as $a\,(>1)$
   increases, the curve gets steeper for
   $x > 0$, and less steep for $x < 0$.

4. 16, 8, 4, 2, 1, 1/2, 1/4, 1/8, 1/16;
$f(x) \to 0$ as $x \to \infty$,
$f(x) \to \infty$ as $x \to -\infty$

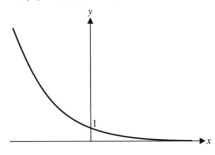

5. If $f(x) = 2^x$, then $(1/2)^x = f(-x)$,
therefore $y = (1/2)^x$ is the reflection of
$y = 2^x$ in the y-axis.

6.

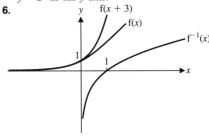

7. (a) $g(x) = 2^x$, $h(x) = 3x - 2$
   (b) 2^{46}
   (c) Stretching by a factor of 3 parallel to
   y-axis and then translating 2 units in
   the direction of the x-axis.

Exercise 19b – p. 297

1. (a) 7.39   (b) 0.368   (c) 4.48
   (d) 0.741

2. (a) $2e^x$ (b) $2x - e^x$ (c) e^x

3. $e^2 - 2$

4. $2 + 2e$

5. 1

6. 0

7. (a)

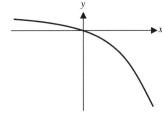

(b)

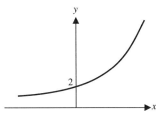

(c)

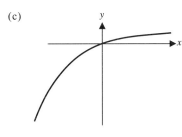

(d)

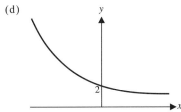

8. (a) $y = 4(2)^{x/6}$
 (c) Any plausible reasons. e.g.
  (i) inherent errors in measuring
   instruments
  (ii) growth rate reduces for $t > 30$
   possible because the dish was left in
   an unsuitable place after this time.

Exercise 19c – p. 300
1. 0.693
2. 0.140
3. −0.693
4. 2.85
5. (a) 1   (b) 2.5   (c) −1   (d) 3

6. (a) 3   (b) 15.8   (c) −0.541
 (d) 1.14
7. (a) $-3e^{-3x}$ (b) $-e^{-x}$
8. (1, 1)
9. (4, ln 4 − 2)
10. (0, 1)
11.

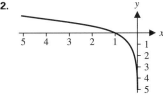

12.

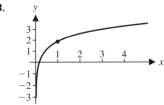

13.

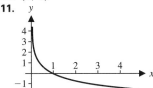

Exercise 19d – p. 302
1. $2e^x + K$
2. $-e^{-x} + K$
3. $\frac{1}{3}e^{3x} + K$
4. $2x - e^x + K$
5. $x + 2e^x + \frac{1}{2}e^{2x} + K$
6. $-e^x - e^{-x} + K$
7. $e^x - 2x - e^{-x} + K$
8. $e^x - e^{-x} + K$
9. $\frac{1}{2}x^2 - 3e^{2x} + K$
10. $2 \ln x + K$
11. $\frac{1}{2} \ln x + K$
12. $\frac{2}{3} \ln x + K$
13. $\frac{1}{2}x^2 - \ln x + K$
14. $x + \ln x + K$
15. $\frac{1}{2}x^2 + 2x + \ln x + K$
16. $2x^2 - 20x + 25 \ln x + K$
17. $\frac{3}{2}x^2 - 2x - \frac{1}{x} + K$

18. $4x - \dfrac{9}{x} - 12 \ln x$

19. (a) $2 \ln 3 - 2 \ln 2$

(b) $\frac{1}{2}e^4 - \frac{5}{2}e^2 + 2e + 1$

(c) $\frac{13}{12} + 2 \ln 4 - 2 \ln 3$

(d) $-\dfrac{1}{e}$ (e) $-\frac{1}{6} + \ln 3 - \ln 2$

(f) $\dfrac{1}{2} - \dfrac{1}{e^2} + \dfrac{1}{2e^4}$ (g) $\frac{1}{2}(e^2 - e^4 + \ln 2)$

(h) $\frac{4}{3} + \frac{3}{2}e^{-2} - \frac{3}{2}e^{-6}$ (i) $\frac{63}{10} - 2e^{1/2}$

20. $1 - \dfrac{1}{e^3}$

21. (b) $\frac{1}{3} + \ln 3$

22. (a) (i) $-\dfrac{1}{e}$ (ii) $e - 2$

(iii) $e - 2 - \dfrac{1}{e}$

(b)

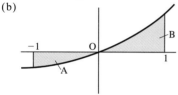

A is (i),   B is (ii),   B–A is (iii)

24. Between -1.5 and -1 and between 0.5 and 1

25. $\ln x$ does not exist for negative values of x.

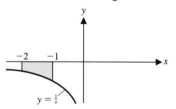

The area shown exists, so $\displaystyle\int_{-1}^{-2} \frac{1}{x}\, dx$ has a value (which is negative).

$\displaystyle\int_{-1}^{-2} \frac{1}{x}\, dx = -\int_{1}^{2} \frac{1}{x}\, dx$

26. Area from $a = 1$ to $a = 6$ is approx 23 units

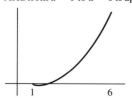

CONSOLIDATION E

Multiple Choice Exercise E – p. 307

1. D
2. B
3. C
4. D
5. A
6. D
7. B
8. D
9. D
10. C
11. C
12. D
13. F
14. T
15. F
16. F
17. F
18. T
19. F
20. T
21. T
22. F

Miscellaneous Exercise E – p. 310

1. (a) 1.65, 65%   (b) 0.5   (d) 1/2
2. (a) 8   (b) -4 (c) 1.72
3. -4, 7
4. 2.218, $x = 2(1 + e^x)$
5. (a) cycles through $0, -1, 0, \ldots$, cycles through 0. $-1, 0 \ldots$ after the first term, diverges.
(b) $1/2 + \sqrt{5}/2$, $1/2 - \sqrt{5}/2$
7. (a) £70 620   (b) 1062
8. $20(1.1)^{n-1}$, 18
9. (a) 29.13   (b) 507.39
10. (a) $|2x| < 1$ (b) $1/(1 - 2x)$
(c) $(1 - (2x)^{10})/(1 - 2x)$
(e) $|x| < 0.315$
11. (a) $2x + 2$ (b) $x^3/3 + x^2 + C$
12. (a) $A(2, 0)$, $B(-2, 32)$
(b) $(-4, 0)$ (c) 108 sq. units
13. (a) $13\frac{1}{2}$
(b)

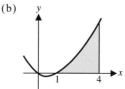

Answers **351**

14. $y = -\frac{1}{12}(16 - 6x + 12x^2 - 8x^3)$

15. (a) $-15/4$ (b) $\pi^4/3$

16. $a = (4n - 1)/2$

17. $a = 1.5,\ b = 6.25$

18.

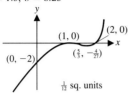

$\frac{1}{12}$ sq. units

19. $x = 1,\ f(x) = 4$ max;
$x = 3,\ f(x) = 0$ min; $6\frac{3}{4}$ sq. units.

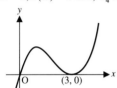

20.

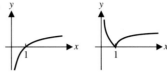

0.14, 7.39
2 (the line $y = 3 - x$ cuts $y = |\ln x|$
in 2 places)

21. (a)

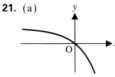

(b) $3 - e^2$

22. $4 + \ln 4$

23. (a) $1 + \ln 2$ (b) $\frac{1}{2}(e^4 - e^2) - 1$

24. (a) $(-1, 2),\ (4, 17)$
(b) $125/6$

INDEX